AF718369

WITHDRAWN

Lubricants and Lubrication in Metalworking Operations

MANUFACTURING ENGINEERING AND MATERIALS PROCESSING

A Series of Reference Books and Textbooks

SERIES EDITORS

Geoffrey Boothroyd

Department of Mechanical Engineering
University of Massachusetts
Amherst, Massachusetts

George E. Dieter

Dean, College of Engineering
University of Maryland
College Park, Maryland

1. Computers in Manufacturing, *U. Rembold, M. Seth and J. S. Weinstein*
2. Cold Rolling of Steel, *William L. Roberts*
3. Strengthening of Ceramics: Treatments, Tests and Design Applications, *Henry P. Kirchner*
4. Metal Forming: The Application of Limit Analysis, *Betzalel Avitzur*
5. Improving Productivity by Classification, Coding, and Data Base Standardization: The Key to Maximizing CAD/CAM and Group Technology, *William F. Hyde*
6. Automatic Assembly, *Geoffrey Boothroyd, Corrado Poli, and Laurence E. Murch*
7. Manufacturing Engineering Processes, *Leo Alting*
8. Modern Ceramic Engineering: Properties, Processing, and Use in Design, *David W. Richardson*
9. Interface Technology for Computer-Controlled Manufacturing Processes, *Ulrich Rembold, Karl Armbruster, and Wolfgang Ülzmann*
10. Hot Rolling of Steel, *William L. Roberts*
11. Adhesives in Manufacturing, *edited by Gerald L. Schneberger*
12. Understanding the Manufacturing Process: Key to Successful CAD/CAM Implementation, *Joseph Harrington, Jr.*
13. Industrial Materials Science and Engineering, *edited by Lawrence E. Murr*
14. Lubricants and Lubrication in Metalworking Operations, *Elliot S. Nachtman and Serope Kalpakjian*

OTHER VOLUMES IN PREPARATION

Lubricants and Lubrication in Metalworking Operations

ELLIOT S. NACHTMAN
Tower Oil & Technology Co.
Chicago, Illinois

SEROPE KALPAKJIAN
Illinois Institute of Technology
Chicago, Illinois

MARCEL DEKKER, INC. New York and Basel

TJ
1077
.N24
1985

Library of Congress Cataloging in Publication Data

Nachtman, Elliot S.
Lubricants and lubrication in metalworking operations.

(Manufacturing engineering and materials processing ; 14)
Includes bibliographies and index.
1. Metal-working lubricants. I. Kalpakjian, Serope.
II. Title. III. Series.
TJ1077.N24 1985 671.3 85-4365
ISBN 0-8247-7401-9

COPYRIGHT © 1985 by MARCEL DEKKER, INC. ALL RIGHTS RESERVED

Neither this book nor any part may be reproduced or transmitted in any form or by any means, electronic or mechanical, including photocopying, microfilming, and recording, or by any information storage and retrieval system, without permission in writing from the publisher.

MARCEL DEKKER, INC.
270 Madison Avenue, New York, New York 10016

Current printing (last digit):
10 9 8 7 6 5 4 3 2 1

PRINTED IN THE UNITED STATES OF AMERICA

TO
EMOLYN
AND
MARI AND VARTKES

Preface

The mechanics of lubrication and the selection, control, application, and disposal of metalworking lubricants influence the effectiveness of metalworking processes and the overall efficiency of the manufacturing operation. An understanding of these, as well as various other aspects of lubrication, is of particular importance as changes in technology and competition demand improved manufacturing productivity, product quality, and cost control.

This book treats metalworking lubrication from the viewpoint of manufacturing operations. It consists of 17 chapters and covers the necessary technical fundamentals, as well as the many operating aspects of metalworking lubrication. The introductory chapters deal with the basic mechanics of metal forming and removal, surface phenomena, and the fundamentals of lubrication. The subjects of friction and wear are discussed in terms of their relevance to lubrication in metalworking. The chemistry and properties of metalworking lubricants are then described, followed by chapters on prescreening, application methods, maintenance, and handling of lubricants. Because of their growing importance, the biological and regulatory aspects of metalworking lubricants are treated in separate chapters. The last two chapters summarize recommended lubricants and practices for metalworking processes.

It is our hope that the text, as written and organized, will provide insight and a source of information for manufacturing plant personnel, such as machine operators, engineers, chemists, biologists, supervisors, and managers. In addition, we hope that students in manufacturing and various other related engineering disciplines, industrial chemistry, and biology will also gain an understanding of the complexity of metalworking lubrication technology.

We are grateful to many colleagues, former students, and teachers, as well as staff members and production workers at various manufacturing plants, for their help and contributions to the contents of this book. We are particularly indebted to many staff members at Tower Oil & Technology Company: Joseph Ivaska, whose suggestions led to the initiation of this work, and who reviewed individual chapters critically and suggested various improvements in the text and Don Dethmers, Norman Henson-Mack, Elroy Spallina, Blanca Mireles, and John Znavor. The review by Roger P. Hunz of Keil Chemical Company of Chapter 15 was very useful. This book could not have been compiled without the skilled assistance of Jean Kalpakjian, who produced copy from chaos with patience and care.

The format of the book and the approach to the subject of metalworking lubrication, along with any errors of commission as well as omission, are the exclusive responsibility of the authors. We would be grateful to the reader if any errors or omissions uncovered in reading this text would be brought to our attention.

Elliot S. Nachtman
Serope Kalpakjian

Contents

Lubricants and Lubrication in Metalworking Operations

1

Introduction:

The Technology of Lubrication in Metalworking Processes

Approximately 60 million gallons of lubricants are used each year in metal cutting and forming operations in the United States, costing approximately $350 million. This cost is very much less than the costs associated with the purchase of tools, machines, materials to be processed, or the relevant labor costs. Expenditures for lubricants, however, have an economic importance that far transcends their cost, because their presence or absence may make the difference between a cost-effective operation and economically unacceptable production.

Our analysis of the technology of metalworking lubrication in this book includes considerations of the mechanics of metal forming and cutting operations, the chemistry of lubricants, the interactions of surfaces of work materials and tooling and the lubricant, the methodology of lubrication, the storage, analysis, and control of lubricants, screening and testing, the biology of lubricants, the effects of laws and regulations, and, finally, a systematic analysis of specific metal cutting and forming operations as influenced by selected lubrication technology.

Lubrication of metalworking operations may be unnecessary or even harmful in a few selected circumstances. We only consider those operations in which lubrication of the interface between the tool and the workpiece is necessary to achieve quality, as well as economical production of the workpiece in question. In the context of the following chapters, metalworking is meant to describe those operations that involve metal removal or metal forming. It is recognized that, frequently, these operations are carried out successively.

The metals and alloys that may be worked have a wide range of possible compositions: steels, brasses, titanium alloys, and aluminum alloys, as well as commercially pure metals, such as copper and aluminum. These metals, furthermore, may have metallic, organic, or inorganic coatings applied prior to the metalworking process. The resulting surface influences the selection as well as the performance of the lubricating film.

Metal removal processes, such as turning, milling, broaching, drilling, tapping, cutoff, grinding, polishing, and lapping, remove metal by applying a tool or an abrasive at some appropriate speed with a force sufficient to remove metal in the form of chips. The tool composition, geometry, and stiffness greatly influence the efficiency of the operation and the surface integrity, finish, and tolerance of the part. Furthermore, the machine tool structure also influences the cutting conditions and the results of the operation.

Metal forming processes, such as forging, rolling, extrusion, rod and wire drawing, tube drawing, deep drawing, swaging, and roll forming, form metal through plastic deformation. The tool and die geometry, surface finish, and stiffness critically influence the mode of plastic flow and hence the shape, surface integrity, finish, and tolerance of the formed part. The lubricant film, in turn, provides a critical additional dimension to such operations.

For a given metalworking process, an approach to optimum lubricant selection depends upon the identification of the process, followed by an analysis of each component of the operation. For example, the identification and analysis of components might consist of consideration of the following factors: the condition of the metal stock (shape and coating), method of application of the metalworking fluid, choice of metalworking equipment (press type, speed, tooling and its composition, surface finish, and geometry), secondary operations (welding and cleaning) and their costs as influenced by lubricant costs, storage conditions (humidity and rust protection), and compliance with regulations (the Occupational Safety and Health Administration, OSHA, and the Environmental Protection Agency, EPA). These are some of the major operational components of a metalworking process that directly influence the lubricant choice. The lubricant will influence the cost of many of these aspects of the operation, as well as the quality of the product.

Since it is not always economically practical to proceed from the liquid metal state to a solid part or shape directly, in most cases intermediate metalworking operations must be carried out. In general, these operations range from hot-working processes involving the direct generation of a shape from an ingot to subsequent finishing of this shape by cold-deformation processes. These processes are often followed by metal removal processes. The economics of the total system involved is specifically dependent upon operating decisions about each of the components of the system. Thus, hot, warm, or cold forming, as well as removal processes carried out on metallic workpieces of various shapes, finishes, and compositions, make up the systems to be discussed in the following chapters.

The analysis of the total metalworking process may be assisted through the use of value analysis methodology. Therefore, we will describe the general background of metalworking processes, followed by the application of value analysis to maximize the economic return of such processes. Our interest in the application of lubricants to metalworking is based on the need for improving the quality of the product and the cost effectiveness of each component of the process.

1.1 DESCRIPTION OF PRIMARY, SECONDARY, AND PARTS-MAKING PROCESSES

Metalworking processes consist of two main classes: metal forming and metal removal. In the transition from the molten metal to a finished part, most frequently there are intermediate stages of, first, formation of an ingot, followed by hot rolling or hot extrusion and, subsequently, cold rolling, drawing, or cold extrusion. Hot-working processes are generally referred to as primary metalworking or bulk-deformation processes. Lubrication of these processes, particularly in extrusion, governs directly whether the process is carried out in a cost-effective manner.

Such operations as cold rolling, drawing, and cold extrusion are generally known as secondary metalworking operations. Liquid lubricants, as well as solid-film-forming lubricants or paste, are the lubricants of choice for such operations.

Subsequently, the parts-making processes come into play. These processes, such as roll forming, heading, press operations of all types, screw machining, drilling, tapping, or some combination of these or other metalworking processes, almost always depend upon the appropriate application of lubricants to permit cost-effective operations.

1.1.1 Materials: Tools and Workpiece

A very broad array of materials is used in fabricating the tool necessary for forming in either the primary, secondary, or parts-making operation. The metal formed also varies widely in composition, from almost pure metals to very complex alloys.

Some tools are made from heat-treated alloy steels hardened to between 60 and 70 HRC. These tools are commonly used for rolling operations, both hot and cold, and deep drawing, as well as various other operations. Carbide tooling, which consists of tungsten carbide or combinations of various carbides bound together by binders, such as cobalt, through powder metallurgical techniques are turned into compacts that have very good high-temperature and wear properties. Carbide tooling is generally used for high-speed metal cutting operations, as well as high-temperature, high-speed metal forming operations, such as Sendzimir rolling, in which large reductions are achieved through the use of multiroll mills.

Finally, ceramic tooling is beginning to have expanded applications in metalworking operations, particularly in metal cutting, but also in wire drawing operations in which the high-temperature service properties are often superior to carbide tooling.

Inappropriate lubrication of the tool may prevent cost-effective metalworking. In rolling, for example, inappropriate lubrication may cause an increase in power requirements, the degree of deformation achievable is reduced, and, in extreme situations, deformation may not take place at all. For example, in rolling, deformation may not take place and the rolls skid in the roll gap.

1.1.2 Cost of Tool Wear: Defective Parts and Downtime

Let us now consider some of the costs related to these operations. Subsequently, we will discuss some of the auxiliary operations that are directly influenced by and that also have an impact on any value analysis of a change in the metalworking process.

In the first place, one must consider the cost of the tooling. If the tooling cost can be reduced, through appropriate lubricant choice, if a change in the geometry of the tool due to tool wear can be delayed, or if chipping or wear of the tool can be reduced, then tool life can be extended.

If the shape produced by the metal forming process is out of tolerance and the surface finish is unacceptable, if there are defects or cracks in the surface or throughout the body of the part, or if the shape is distorted and not acceptable, the cost of the process is clearly unacceptable.

If should be pointed out that machine downtime is one of the most pernicious of manufacturing costs. Maximizing uptime and reducing downtime can profoundly affect the economics of the process. Appropriate lubrication can be a critical component of the manufacturing process designed to maximize uptime.

1.1.3 Auxiliary Process Cost Implications

A manufacturing decision on whether to use a recycling system for the lubricants can significantly affect the costs of the operation. It may be economical to develop a once-through lubrication system with no recycling. On the other hand, if recycling appears to be economical, the total cost of recycling must then be determined. Control of the fluid composition, control of contamination, and disposal of waste fluid are all considerations involved in any value analysis of lubrication.

If welding or painting is to follow the parts-making operation, then the impact of the lubricant film on these subsequent operations must be evaluated. If cleaning is required prior to the painting or welding operation, then the cost of cleaning in terms of energy, disposal, and operation must be evaluated. Selection of lubricants that do not require removal is obviously an alternative to be investigated. If a series of metalworking operations involving either metal forming or cutting is required, then the interrelationship of the lubricant in each of the steps should be evaluated. Will succeeding applications of lubricants be required? Will the same lubricant be suitable for the subsequent operation? Are they compatible with each other?

The method of application of the lubricant may also be important. Will the lubricant be applied by spraying or roller coating? Should it be applied over the whole surface of the material to be formed, or should it be applied only to the tooling? These process cost questions serve to illustrate the importance of identifying and analyzing both the process and auxiliary operating costs.

1.1.4 Productivity: Competitive Processes and Alternative Methods

Selection of the most cost effective process or combination of processes required for the production of finished parts is among the most important manufacturing decisions. To use a process that removes metal and incurs the cost of the scrap and its disposal, or to use a metal forming process that may require compromises on shape and tolerance, is fundamental to the value analysis process, which will now be briefly discussed. It is very important, in making the manufacturing decision, that all parts of the competing systems be considered, in terms of their intrinsic cost benefits as well as the interrelationship of each of the components of the system. It is with respect to a methodology of analysis, which has been proven valuable in attacking these problems, that we now turn to a brief consideration of value analysis.

1.2 VALUE ANALYSIS

A systematic procedure for evaluating and analyzing change as it may result in equal or better performance of a system at a lower total cost is known as value analysis. The total cost is considered to be the sum of all work and expenditures made in the development, production, and application of each component of a manufacturing system. The technique of value analysis may be considered as consisting of a series of steps to be carried out.

1. Initiation. Establish the necessary working relationships and methodology for agreeing on value and securing concrete data and information.
2. Cooperation. A value analysis is useless if it does not result in action. Cooperation of all parties – engineers, operators, suppliers, managers, purchasing agents, quality control personnel, and union members – can contribute to successful change. Lack of cooperation can make change difficult, if not impossible; therefore, establish cooperation.
3. Information. Secure facts; determine all costs and relationships.
4. Definition. Define the function of each of the components of the existing system, and then determine the value of their functional relationships.
5. Creativity. Analyze relationships, establish positive thinking, and create new relationships, ideas, and concepts to improve the cost-benefit ratio.
6. Evaluation. Define and combine ideas, establish values, present costs, and motivate a positive response to the information.
7. Investigation. Refine ideas into workable and salable solutions to provide lower cost methods of performing the desired functions. Use company and industrial standards; consult vendors and specialists.
8. Recommendation. Present the facts, costs, and values, and motivate positive action.

Systematic analysis of metalworking lubrication will result in the establishment of agreed-upon costs and values fundamental to cost-effective production.

Letus consider the following question and apply value analysis to illustrate the use of this technique. Should we substitute a water-soluble metal-working fluid for a drawing oil in deep drawing of plain-carbon steel parts that must be painted prior to assembly? The raw material is purchased in the form of strip.

1. Initiation. The important areas on which selection of the fluid may have some impact are: control of the fluid, corrosion due to the water content of the water-soluble fluid, cleaning costs, attitude of the employees, corrosion of tooling, and disposal.

2. Cooperation. Identify personnel involved in the question being studied: operators, management, engineering, quality control, and others. Determine relationships, informal and formal. Investigate cost study procedures, union position, and customer needs and their relationship to relevant operational questions.
3. Information. Carbon steel strip is coated with a slushing oil to prevent rust during transportation from the source of supply to the manufacturer. What effect will this slushing oil have on the metalworking lubricant? How is the lubricant applied at present? Is the lubricant recirculated? Is there a problem of lubricant buildup on the press, and how often is the tooling changed because of wear? Are the parts to be cleaned after the drawing operation? What is the cost of cleaning (temperature required and cost of cleaning chemicals and their disposal)? Who will ensure that the fluid has the appropriate concentration and is kept clean? Is the present lubrication system adequate?
4. Definition. Lubricant must permit accurate forming of the workpiece with no wrinkling or cracking in the walls. Secondary functions are that it successfully operate through the coating on the strip from the mill supplier and that it be easily removed during the cleaning operations. It should not cause dermatitis, and it should not contribute to rusting of the press, tooling, or the parts.
5. Creativity. Can we select a lubricant that would eliminate the cleaning operation? Can the application method be made more efficient? Can the lubricant film be designed to permit painting directly over its film, with improved paint adhesion?
6. Evaluation. Selection of an appropriate water-based fluid that can be painted is a possibility. Can such a fluid successfully provide lubrication during the forming process? Will it clean off the rust preventative provided at the mill? What are the relative costs of each of these effects and attributes? What is the cost of controlling a water-soluble lubricant? Is mixing done under proper control?
7. Investigation. Alternative fluids must be evaluated with respect to interaction with the corrosion protective film provided at the hot mill. Determine the cost of control of the fluid relative to the control of the previously used oil-based lubricant, effects on tool life, part integrity, corrosion, and dermatitis, and, finally, the ability of the paint line to adequately coat the part without requiring the removal of the fluid. Methods of application must be evaluated. Tool geometry, surface finish, and lubricant interrelationships need to be investigated.

8. Recommendation. The information that has been collected as a result of the value analysis process is analyzed and presented. The recommended practice is defined in terms of costs and benefits. Credit to those who participated in the study is specifically given and distributed throughout the organization. Identification of those personnel who are willing to adopt new ideas and new procedures is a critical aspect of the successful application of value analysis.

BIBLIOGRAPHY

Mason, Jr., J. A., Cost savings through cutting fluid selection, 1969, ASTME Technical Paper MR 69-259.

Principles and Applications of Value Engineering, 1981. Defense Management Joint Course, 7th printing, May 1981, U.S. Government Printing Office.

Report on U.S. Lubricating Oil Sales, 1982. National Petroleum Refiners Association, Washington, D.C.

Valentine, R. F., 1970. *Value Analysis for Better Systems and Procedures,* Prentice-Hall, Englewood Cliffs, New Jersey.

2
The Mechanics of Lubrication in Metal Forming

A metal forming operation consists of a machine, tooling, lubricant, and the workpiece. A detailed analysis of this system reveals a large number of parameters, often interacting with each other. A factor of fundamental importance is the friction at the interface between the tool and the workpiece. Friction influences tool life, part surface finish, metal flow, forces, and power required and may lead to lubricant degradation. Since friction at the tool-workpiece interface is influenced by many parameters, including the lubricant (see Chap. 5), each metal forming system must be studied individually.

The behavior of an interface between a tool and workpiece depends on such parameters as the physical, mechanical, and metallurgical characteristics of the contacting bodies. In addition, the process parameters play a role, such as the relative motion and speed at the interface, the magnitude of contact stresses, and the temperature of the bodies in contact. Other factors can also have an influence, such as the dynamic characteristics of the metalworking machine.

In this chapter the mechanics of major metal forming processes are briefly reviewed with respect to the effects of friction and its control. Significant process parameters are outlined as they affect metal flow, forces and stresses, energy requirements, temperatures, wear of tools and dies, and product integrity. Some formulas are given to indicate the effect of friction on forming forces, stresses, and die pressures. It is beyond the scope of this book to describe in detail their derivations or to outline various analytical techniques that are being developed to obtain more precise formulas. More detailed information is available in the Bibliography at the end of this chapter. The physical, chemical, and metallurgical nature of surfaces is described in Chap. 4, and the mechanics of friction, wear, and lubrication is discussed in Chap. 5.

2.1 BULK-DEFORMATION PROCESSES (PRIMARY WORKING)

In this category are such processes as forging, rolling, extrusion, rod and wire drawing, and seamless tube making. These primary working operations are characterized by relatively low ratios of surface area to volume of the workpiece. Sheet-metal forming is an example of a secondary metalworking operation and is characterized by relatively high surface area-volume ratios. Because friction occurs at the tool-workpiece interfaces, the workpiece surface area-volume ratio is important in characterizing the metalworking operation.

2.1.1 Forging

Forging is a process in which the workpiece is basically subjected to compressive forces to plastically deform the material, to make it flow in certain directions. Although the variety in forging geometry and size is very extensive, the mechanics of the process can quite appropriately be studied by analyzing the forging of a simple cylindrical workpiece, such as that cut from a round bar, which is then subjected to compressive forces between two flat platens or dies (Fig. 2.1). As the height of the specimen is reduced, the material flows radially outward since its volume must remain constant. Because the material is in compression the relative motion at the interfaces, due to the radial expansion of the specimen, leads to frictional forces. These forces act radially

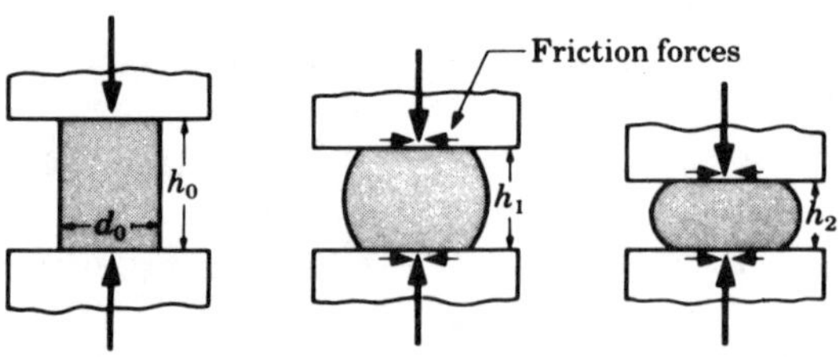

Figure 2.1 Forging (upsetting) of a cylindrical workpiece between flat platens showing barreling due to friction at die-workpiece interfaces. (S. Kalpakjian, 1984, *Manufacturing Processes for Engineering Materials*, Addison-Wesley, Reading, Massachusetts, p. 290.)

inward and cause barreling of the specimen because the surfaces are restrained from flowing outward freely. As friction increases, the barreling increases.

The presence of friction at the die-workpiece interfaces has two significant effects. First, it causes frictional heating, raising the temperature at the interface. This, in turn, may affect the properties and surface finish of the material at or near the surfaces. Also, if present, the lubricant properties and behavior may in turn be affected by this temperature rise.

The second effect of friction is to change the state of stress on the material, which in turn affects the forging load. There are various methods of calculating these stresses. Using a simple method of analysis (known as the slab method, which uses yield criteria to correlate stresses in three principal directions), it can be shown that in simple upsetting the normal stress distribution at the interface is given by the expression

$$p = Ye^{2\mu(a-r)/h} \qquad (2.1)$$

where Y is the yield stress of the material, μ is the coefficient of friction, a is the radius, r is any radius, and h is the height. This formula indicates that the compressive stress, that is, die pressure, increases from the outer radius a toward the center of the workpiece. This pressure distribution is known as the "friction hill" because of its particular shape. It can be seen that friction contributes very significantly to the forging load, because without friction the pressure distribution would be uniform, as can be seen by letting the friction coefficient be zero in Eq. (2.1). The load F for simple upsetting can be given approximately as

$$F = \pi a^2 Y \left[1 + \frac{2\mu a}{3h}\right] \qquad (2.2)$$

For strain-hardening materials Y should be an average yield stress of the material.

The role of the friction coefficient is clearly evident in these expressions. It increases forces and hence energy requirements; it leads to temperature rise, and if excessive, it can lead to surface damage. Relative motion and friction also contribute to die wear, as discussed in Chap. 5. Because of barreling and similar constraints to free flow of the material, the workpiece also undergoes nonuniform deformation throughout its bulk, resulting in nonuniform properties, as well as preferred alignment of impurities and inclusions resulting in mechanical fibering in the forged part. The role of friction at die-workpiece interfaces should not, however, always be regarded as detrimental. Because of its restraining effect, friction can be a very useful means of controlling the flow of the material in more complex forgings.

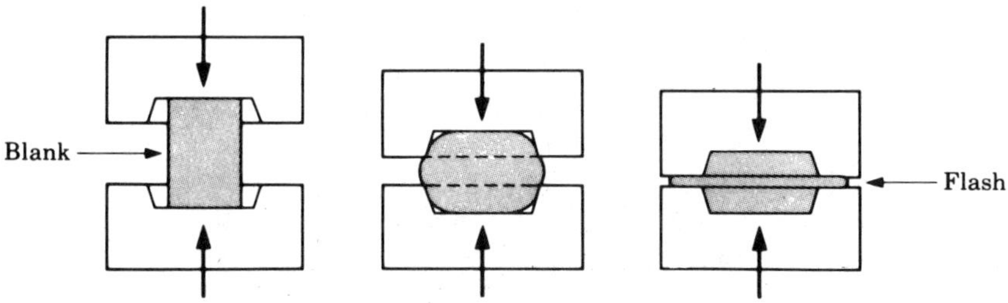

Figure 2.2 Die filling in closed-die forging showing flash formation (S. Kalpakjian, 1984. *Manufacturing Processes for Engineering Materials,* Addison-Wesley, Reading, Massachusetts, p. 298.)

To minimize the work of deformation, the material flows in the direction of least resistance. This simple principle has been used extensively in die design, the most common example the flash in closed-die forging. Figure 2.2 shows an example in which, if the material is not restrained from flowing laterally outward, it will not fill the die cavity properly. In such complex operations, the design of dies and the proper control of friction through the use of various lubricants must therefore be coordinated.

Numerous other examples can be given to indicate the importance of the control of flow of the material in the die cavity by such means. It should also be pointed out that, for forging at elevated temperatures, there are additional problems of the part being chilled when in contact with relatively cool dies. Because of the large variation of temperature along all die-workpiece interfaces and the response of the material to high temperature, and because of the role of the lubricant with its lubricating as well as thermal barrier characteristics, material flow control can be a very complex problem in bulk-deformation processes.

In addition to the foregoing considerations, another significant process variable is the rate at which the forging operation is carried out. A hydraulic press, for instance, generally deforms the material at a slower speed than a mechanical press or a drop hammer. This, in turn, changes the time of contact between the die and workpiece interfaces. The resulting variations in temperature and friction will have an effect on the forging load and material flow pattern. An additional factor is the strain-rate sensitivity of the workpiece material, which reflects that engineering metals and alloys exhibit increasing resistance to deformation at elevated temperatures as the rate of deformation increases.

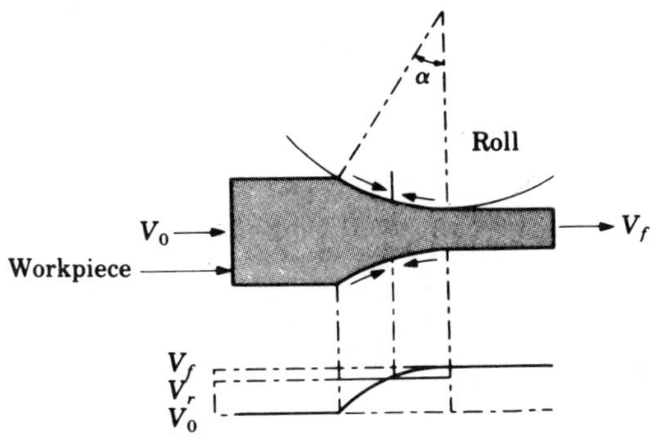

Figure 2.3 Rolling of a strip showing relative velocities between the roll and strip. The two velocities are the same at the netural point. (S. Kalpakjian, 1984. *Manufacturing Processes for Engineering Materials,* Addison-Wesley, Reading, Massachusetts, p. 333.)

2.1.2 Rolling

In this process a strip of metal is reduced in thickness continually by passing it through a set of rolls (Fig. 2.3). In general, the strip is much wider compared with its thickness, and hence the process may be regarded as one of plane-strain deformation. In other words, the reduction in thickness results in an increase in the length of the strip, with little or no change in its width during rolling. To maintain mass continuity, the material in the roll gap travels at an increasingly higher speed as its thickness is reduced. Because the roll is rigid, there is relative motion between the surface of the strip and the roll. There will, however, be one point at which they both travel at the same speed. This point is known as the neutral point or neutral line. Much as in the example of forging in Fig. 2.1, this situation results in a friction hill with a roll pressure distribution, as shown in Fig. 2.4. The area under this curve is simply the roll separating force per unit width of the strip.

Using the slab method of analysis or similar methods, it can be shown that the pressure distribution in the roll gap is given by

$$p = Y' \frac{h}{h_o} e^{\mu(H_i - H)} \tag{2.3}$$

for the entry zone and by

$$p = Y' \frac{h}{h_f} e^{\mu H} \tag{2.4}$$

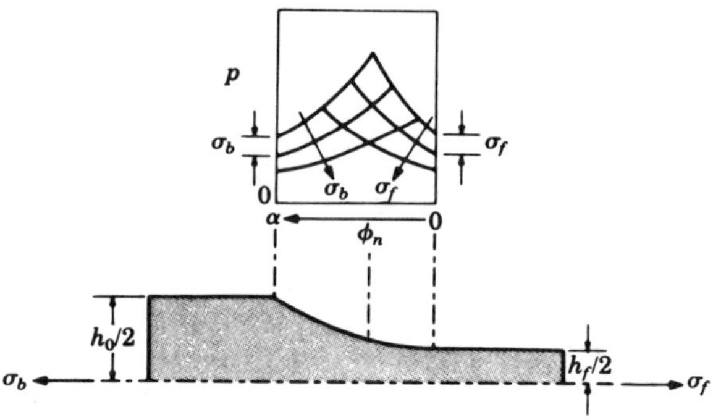

Figure 2.4 Roll pressure distribution showing the effect of front (σ_f) and back tensions (σ_b) on pressure. (S. Kalpakjian, 1984. *Manufacturing Processes for Engineering Materials,* Addison-Wesley, Reading, Massachusetts, p. 339.)

for the exit zone, where p is the contact pressure at a particular location in the roll gap, Y′ is the yield stress of the material in plane strain, and h is the strip thickness at any location in the roll gap. The subscripts o and f indicate the initial and final thickness of the strip, respectively, μ is the coefficient of friction, and H is a convenient parameter defined as

$$H = 2\sqrt{\frac{R}{h_f}}\tan^{-1}\left(\sqrt{\frac{R}{h_f}}\,\phi\right) \tag{2.5}$$

where R is the roll radius and ϕ is the angular position in the roll gap measured from the vertical.

It can be seen from these expressions that as friction increases the roll pressure increases, thus increasing the roll separating force and the power required. Without friction, however, rolling with powered rolls cannot be carried out since the rolls will simply slip. (Rolling can also be done by pulling the strip through idling rolls, known as a Steckel mill.) The friction coefficient can, of course, be varied by using different lubricants and by changes in the roll surface finish. The effects of friction and thickness reduction cause nonuniform deformation of the strip, thus producing residual stresses in the rolled strip.

Rolling may be carried out at room temperature or at elevated temperatures, depending on the ductility of the material, its strength and thickness, and the type of equipment available. Proper selection of lubricants can in-

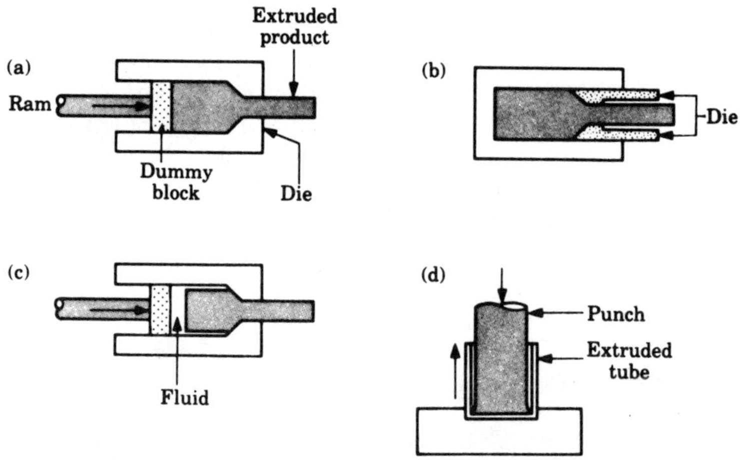

Figure 2.5 Basic methods of extrusion: (a) direct, (b) indirect, (c) hydrostatic, (d) impact extrusion. (S. Kalpakjian, 1984. *Manufacturing Processes for Engineering Materials,* Addison-Wesley, Reading, Massachusetts, p. 351.)

fluence not only friction but also the properties of the surface layers of the rolled metal, its oxide layer thickness, the residual stress level, and surface finish. Oxidation and the presence of oxide layers can become a significant problem in hot rolling, affecting surface finish and integrity.

2.1.3 Extrusion

The extrusion process involves reduction in the cross-sectional area of the workpiece by forcing it through a converging die. In direct extrusion (Fig. 2.5a), the billet travels toward the die opening, all surfaces sliding along the container and die walls. This relative motion under pressure generates frictional forces, thus contributing to the increase in the extrusion force. This can be somewhat avoided in the indirect extrusion process (Fig. 2.5b), in which the billet is stationary with respect to the container, although there is still relative motion, hence friction, between the die and the workpiece material in the die zone.

An approximate expression for the extrusion pressure p in direct extrusion is given by

$$p = \bar{Y}(0.8 + 1.2 \ln R) + \frac{4kL}{D_0} \tag{2.6}$$

where $\bar{Y}$ is the mean flow stress of the material, R is the extrusion ratio (the ratio of the original and final cross-sectional areas), L is the billet length, D_0 is the billet diameter, and k is the shear stress at the billet-container interface. An upper limit for k is Y/2, where Y is the yield stress of the material.

The extrusion pressure, hence, the extrusion force, is thus a function of the billet length because the longer the billet is in the container, the greater the apparent interface between the billet and the container, and hence the greater the frictional forces to be overcome. When friction is zero, Eq. (2.6) reduces to a minimum, which is the ideal work of plastic deformation, including the redundant work due to the nonuniform deformation of the material in the die region.

Because extrusion forces can be quite high, and partly to improve the ductility of the workpiece material, most extrusions are generally carried out at elevated temperatures. Proper lubrication thus becomes essential, particularly in the die region, where the die surfaces are subjected to high stresses, temperatures, and relative motion – factors that all contribute to wear. Temperature rise in extrusion is due to the work of plastic deformation and friction. Almost all the work of deformation, including that due to friction, is converted into heat. If the extrusion speed is high, making the deformation process approach an adiabatic state, then the heat generated increases the temperature of the extrusion substantially. In the extreme case, the temperature can be so high that incipient melting can take place within the workpiece.

In addition to the problems just discussed, extruded products can develop internal defects known as chevron, centerburst, or arrowhead cracks. These are cracks originating at the center of the extruded product and are due to the nonuniform flow of the material in the die zone. It has been shown that the important parameters that control these defects are the die angle, reduction (extrusion ratio), and the presence of impurities or nonmetallic inclusions in the billet. High die angles, low reductions, and poor quality of material may result in defective extrusions. Friction can also have an effect on defect formation.

Another method of extrusion is hydrostatic extrusion (Fig. 2.5c), in which the required forces are transmitted through a fluid medium. Although limited in its applications, this process is particularly suitable for less ductile materials because of the well-known beneficial effect of hydrostatic pressure on the ductility of materials. The fluid medium in this process also serves as a continuous supply of lubricant at the billet-die interface.

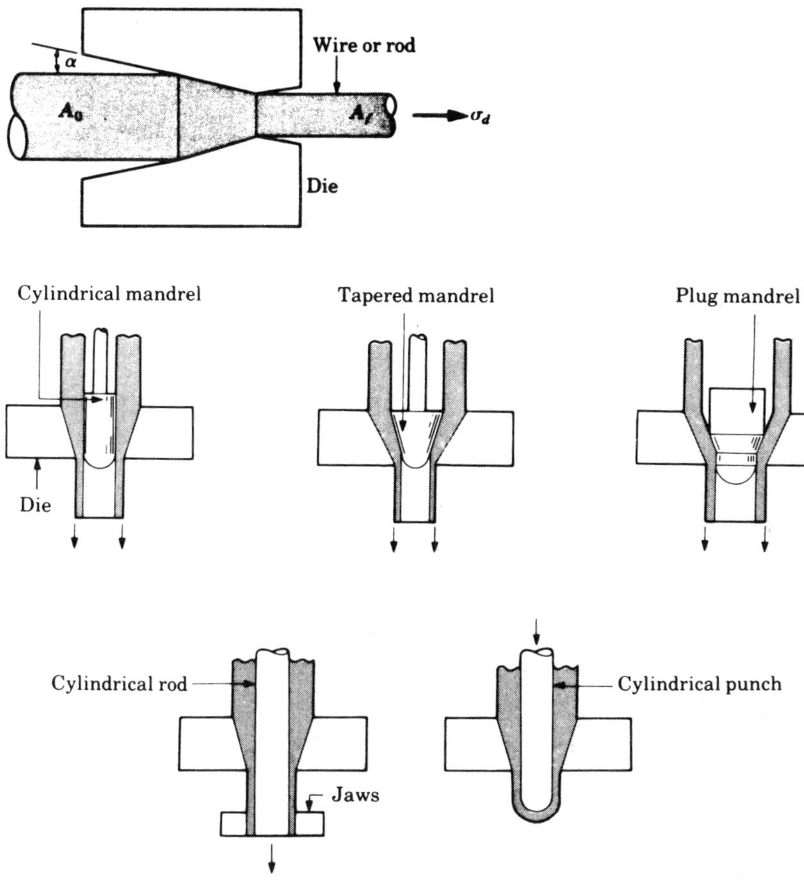

Figure 2.6 Examples of drawing of rod, wire, and tube. (S. Kalpakjian, 1984. *Manufacturing Processes for Engineering Materials,* Addison-Wesley, Reading, Massachusetts, pp. 368 and 376.)

2.1.4 Rod and Wire Drawing

In these processes the material is reduced in its cross-sectional area by pulling it through a die (Fig. 2.6). The die opening may have various shapes, in addition to being round. A wide range of cross-sectional geometries can thus be drawn. Because of the relative motion there is friction at the die-workpiece interface, which increases the drawing force since energy must be supplied to overcome frictional work. The interface is also subjected to normal compressive stresses, that is, die pressure. Analytical studies have shown that the drawing stress σ_d for round bars or wire can be expressed approximately by

$$\sigma_d = \bar{Y}\left(1 + \frac{\tan\alpha}{\mu}\right)\left[1 - \left(\frac{A_f}{A_o}\right)^{\mu \cot\alpha}\right] + \frac{2}{3}\bar{Y}\alpha^2\left(\frac{1-r}{r}\right) \tag{2.7}$$

where α is the die angle, A_f and A_o are the final and original areas of the wire, respectively, r is the reduction of area in drawing and $\bar{Y}$ is the average flow stress of the material.

The role of friction can clearly be seen in these expressions. As in all metalworking processes, the difficulties involved in estimating accurately the friction coefficient make the reliable use of these formulas somewhat questionable. Such formulas, however, do show the effect of friction on forces and die pressures, and hence they are significant in assessing the effects of lubricants.

An approximate expression for the pressure p at any location on the die surface is

$$p = \bar{Y}\left\{1 - \frac{1+B}{B}\left[1 - \left(\frac{A}{A_o}\right)^{B}\right]\right\} \tag{2.8}$$

where $B = \mu/\tan\alpha$ and A is the instantaneous cross-sectional area of the wire in the die zone. It can be seen that the pressure is highest at the die entry and lowest at the exit.

The type of internal cracking observed in extrusion (chevron cracking) can also be found in drawn rods and wire. The relevant parameters are the same as in extrusion, except that higher friction has been shown to increase the tendency for crack formation.

As in all metalworking processes, the work of deformation and friction in drawing can result in a significant rise in temperature and this, in turn, can affect die life and product quality.

2.1.5 Tube Drawing

Tube drawing involves a mechanism similar to that in rod drawing, with the exception that the inside surface of the tube may involve the use of mandrels of various geometries for control of wall thickness and tube size. Figure 2.6 shows various tube drawing processes. The analysis of these processes is similar to that for rod drawing, with friction having, as always, a significant influence on drawing stresses and die pressures. Because of the difficulties of accessibility, lubrication of the plugs and mandrels in tube drawing can be an additional problem, requiring special methods to ensure proper lubrication.

2.2 SECONDARY FORMING PROCESSES

These processes involve a wide range of part shapes and have the characteristic feature of utilizing workpiece materials with high surface area-volume, or thickness, ratios. In sheet forming operations, the material is generally subjected to biaxial stretching in the plane of the sheet using a great variety of tools and dies. Any compressive stresses in sheet forming are due to contact with the dies through which forces must be transmitted. (These are compressive stresses, that is, pressure, normal to the plane of the sheet.) Any excessive compressive stress in the plane of the sheet will cause buckling, wrinkling, and folding of unsupported regions in sheet forming operations.

Sheet forming processes can be classified in various ways, depending on the severity of forming and depth of the part. In addition to stretching, another deformation mode most sheet forming processes involve is bending, along either a straight line or a curve. Of all sheet-metal forming processes, one operation that best exemplifies the problems involved with friction and the mechanics of lubrication is deep drawing.

2.2.1 Deep Drawing

In this process the sheet metal is drawn into a die cavity with the use of a punch (Fig. 2.7). It can be seen that there are many interfaces between the punch, die, and surfaces of the sheet. Before the punch descends, the sheet blank is held in place by a blank holder, the purpose of which is to ensure that the blank does not wrinkle while being drawn into the die cavity. There are thus two interfaces (zone 1 in the figure) in this region: the die-sheet blank, and the blank holder-sheet blank. During drawing, the sheet metal slides radially inward into the die cavity, to be formed into a cup.

Another interface in deep drawing is the die radius-sheet blank region (zone 2), where the bottom surface of the blank continuously slides along the radius R until the end of the punch stroke. The cup being formed is in contact with the punch wall; thus, if there is high friction at this interface (zone 3), the longitudinal tensile stresses in the cup wall (which transmit the drawing force to the rest of the sheet) will be reduced, since the frictional force will support the cup. The interface of the punch head and the blank (zone 4) does not undergo deformation; it does, however, trap some lubricant from the beginning of the punch stroke. This zone can serve as a reservoir of lubricant to be distributed along zone 3 as the punch travels.

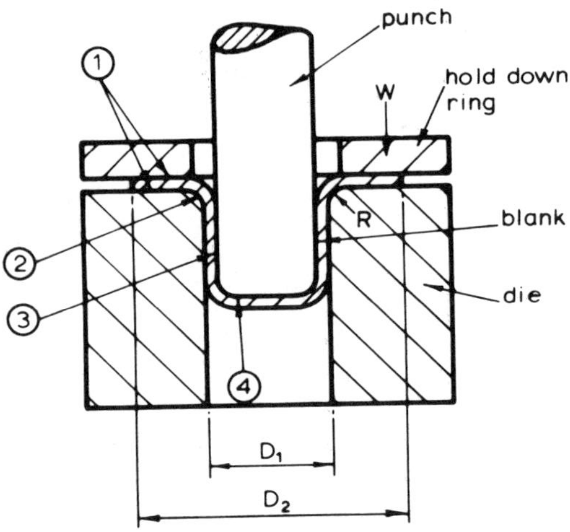

Figure 2.7 Deep drawing a cylindrical cup, showing various zones of lubrication. (W. R. D. Wilson, 1978. In *Mechanics of Sheet Metal Forming,* Plenum Press, New York, p. 168.)

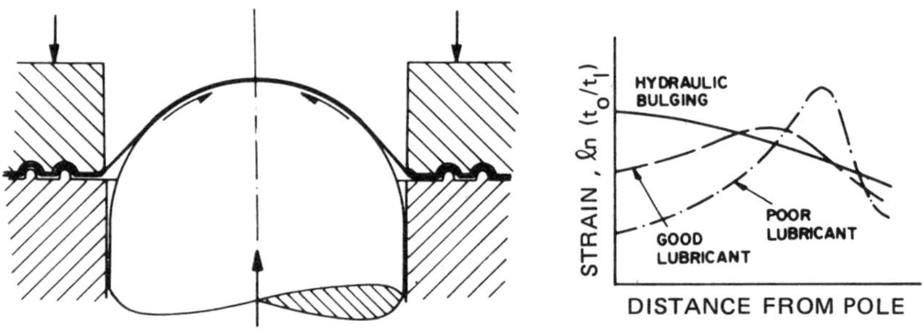

Figure 2.8 Effect of lubrication on the extent and distribution of thinning of sheet metal in bulging hydraulically and with a hemispherical punch. (J. A. Schey, 1983. *Tribology in Metalworking,* American Society for Metals, Metals Park, Ohio, p. 520.)

Although different types of failures are associated with deep drawing and other sheet forming processes, those related to friction and lubrication are punching out the bottom of the cup and localized thinning, as shown in Fig. 2.8. Other types of failures may also be indirectly related to the lack of con-

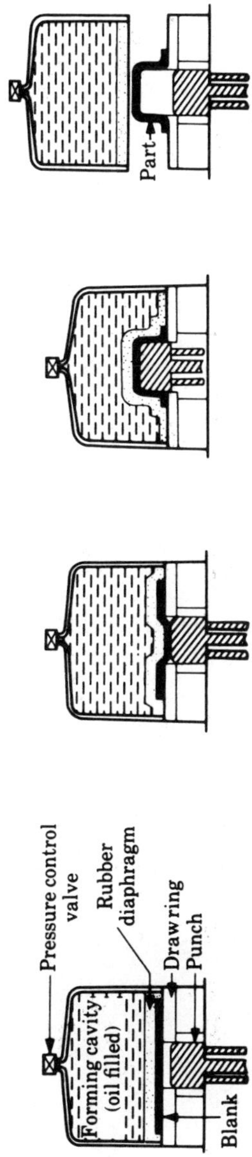

Figure 2.9 Schematic illustration of the hydroform process. (S. Kalpakjian, 1984. *Manufacturing Processes for Engineering Materials*, Addison-Wesley, Reading, Massachusetts, p. 441.)

trol of friction and lubrication. Ideally, it would be desirable to have higher friction between the cup and the punch (zone 3) and lower or no friction between all other interfaces in the drawing operation. In fact, the higher drawing ratios, that is, deeper cups, obtained by the hydroform process (Fig. 2.9) have been attributed to the cup wall being pressed against the punch by the dome pressure, thus increasing the frictional resistance in zone 3 and reducing longitudinal tensile stresses in the cup wall.

The friction at the punch-cup interface must not, however, be high enough to interfere with removal of the cup from the punch. One basic problem is that the portions of the blank that eventually become the cup wall are drawn in from the die-blank holder region (zone 1), where lubrication is desirable. Thus, some of the lubricant is drawn into the punch-cup wall interface, thus reducing the beneficial effect of friction in zone 3.

It is difficult to correctly estimate the punch force P in deep drawing. One expression, which shows the effect of friction, is given by

$$P = \left(\pi D_1 t Y' \ln \frac{D_2}{D_1} + \mu\, 2F_B \frac{D_1}{D_2}\right) e^{\mu\pi/2} + F_R \tag{2.9}$$

where D_2 and D_1 are the blank and punch diameters, respectively, t is the wall thickness, F_B is the blank holder force, and F_R is the force required to bend and unbend the sheet around the die radius.

As for the blank holder pressure required in practice, a generally accepted value is about 1% of the sum of the yield and tensile strengths of the sheet metal. If the blank holder force is too low, the blank may begin to wrinkle during drawing in, and if too high, it increases the punch force (due to the increasing friction force at the die-blank holder region), which in turn increases the possibility of punching out the bottom of the cup during drawing.

From the foregoing discussions it is apparent that successful sheet-metal forming requires careful consideration of all the parameters that play a role in the mechanics of the process. Drawability and sheet-metal formability depend, in general, not only on process parameters but also on the properties and texture of the sheet metal, particularly its anisotropy, both in the planar and in the direction normal to the sheet. However, because these material parameters are not directly influenced by friction and lubrication, it is beyond the scope of this book to discuss them in detail. The Bibliography at the end of this chapter is a helpful guide for further details on sheet forming processes.

Although any changes in the sheet thickness in forming processes is generally due to biaxial stretching of the material, one process in which this is done deliberately is ironing. After a cup is formed, its wall thickness can further be reduced by pushing the cup through an ironing die with the use of a punch, a process somewhat similar to tube wall reduction by drawing. Because the material is undergoing deformation in the ironing die, resulting in a change in length, there are frictional forces not only on the outer wall but also on the inside wall of the cup. Careful control of these forces is essential for successful ironing, particularly for thick-walled cups. Here, the frictional force can be important because of its relatively high magnitude with respect to the various other stresses in the region.

2.2.2 Various Secondary Forming Processes

There is a great variety of sheet forming processes, but the mechanics of deformation is basically similar to that for deep drawing just discussed. Some of the major processes are press brake forming, stretch forming, roll forming, rubber pad forming, and spinning. In some of these processes the relative motion between the tools and dies and the sheet metal is very high, and in others it is very low. Furthermore, the compressive stress normal to the surface of the sheet (die pressure) can have a wide range, which in turn can have a significant effect in friction and lubrication. When large contact surfaces are involved, the uniformity of the lubricant layer can be important for both dimensional control of the part and its surface quality. Relevant details are discussed in later chapters.

2.3 SHEARING SHEET METAL: PUNCHING (PIERCING) AND RELATED PROCESSES

Before a piece of sheet metal is formed by any of the processes, it generally must be cut from a larger sheet or blank. This is done by the basic process of shearing the metal between two dies that subject the material to shearing stresses. Figure 2.10 shows the basic process of shearing, in which we note the relative motion of the punch with respect to the sheet metal and the lower die. Friction at these interfaces can have a significant effect on the quality of the sheared surfaces. It can also contribute to the magnitude of the punch force and to the wear of punches and dies. Studies in shearing have generally concentrated on the effect of clearance and die design on the quality of the sheared edge. It is well recognized, however, that lubrication is important, particularly in fine blanking (Fig. 2.10). Effective lubricants can reduce die wear considerably, in addition to improving the quality and surface finish of the sheared edge. Furthermore, the punch may have the tendency to stick to the workpiece (hang-up) on the return stroke if not lubricated properly.

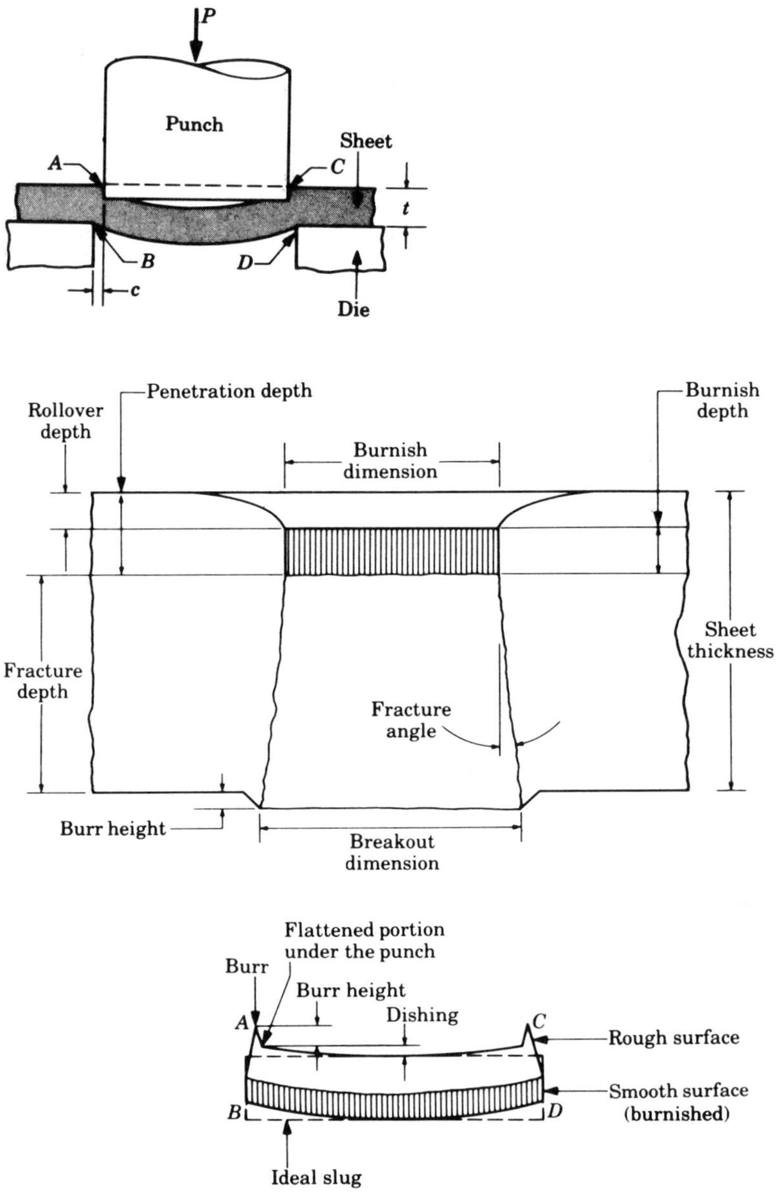

Figure 2.10a Schematic illustration of the shearing process for sheet metal: ordinary shearing. (S. Kalpakjian, 1984.)

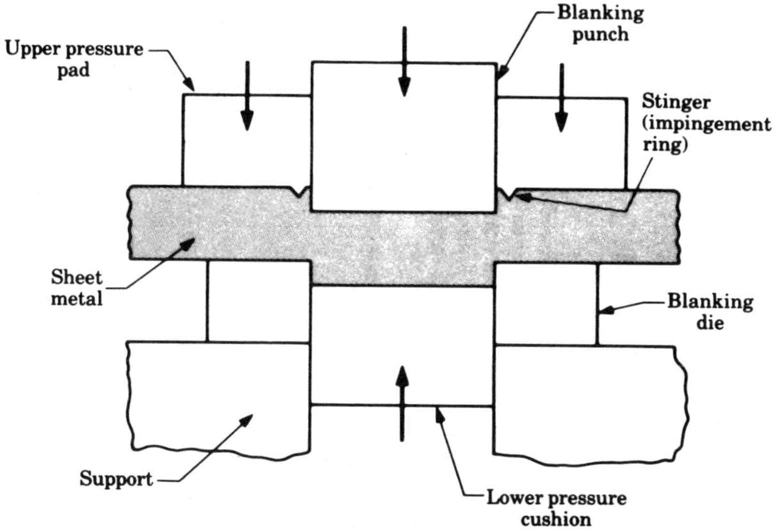

Figure 2.10b Schematic illustration of the shearing process for sheet metal: fine blanking. (S. Kalpakjian, 1984.)

From the foregoing discussions it can be seen that friction and lubrication have very significant effects on material flow characteristics, product integrity, and the forces and pressures involved in all metal forming processes.

The proper selection and application of lubricants in controlling friction thus become self-evident. Various details of friction and lubrication mechanisms are discussed in Chapter 5 as they relate to metalworking operations.

BIBLIOGRAPHY

Altan, T., F. W. Boulger, J. R. Becker, N. Akgerman and H. J. Henning, 1973. *Forging: Equipment, Materials and Practices,* Battelle Memorial Institute, Columbus, Ohio.

Bartz, W. J., and J. Wolff, eds., 1982. *Lubrication in Metal Working,* Technische Akademie Esslingen, Germany.

Eary, D. F., and E. A. Reed, 1974. *Techniques for Pressworking Sheet Metal,* 2d ed., Prentice-Hall, Englewood Cliffs, New Jersey.

Everhart, J. J., 1964. *Impact and Cold Extrusion of Metals,* Chemical Publ. Co., New York.

Gannon, H. E., ed., 1982. *The Making, Shaping and Treating of Steel,* 10th ed., U.S. Steel Corp., Pittsburgh.

Kalpakjian, S., 1984. *Manufacturing Processes for Engineering Materials,* Addison-Wesley, Reading, Massachusetts.

Koistinen, D. P., and N. -M. Wang, eds., 1978. *Mechanics of Sheet Metal Forming,* Plenum Press, New York.

Larke, E. C., 1963. *The Rolling of Strip, Sheet, and Plate,* 2d ed., Chapman and Hall, London.

Laue, K., and H. Stenger, 1981. *Extrusion,* American Society for Metals, Metals Park, Ohio.

Metals Handbook, 8th ed., Vol. 4, *Forming,* 1969. American Society for Metal, Metals Park, Ohio.

Metals Handbook, 8th ed., Vol. 5, *Forging and Casting,* 1970. American Society for Metals, Metals Park, Ohio.

Pearson, C. E., and R. N. Parkins, 1961. *The Extrusion of Metals,* 2d ed., Wiley, London.

Roberts, W. L., 1978. *Cold Rolling of Steel,* Marcel Dekker, New York.

Roberts, W. L., 1983. *Hot Rolling of Steel,* Marcel Dekker, New York.

Sachs, G. and H. E. Voegeli, 1960. *Principles and Methods of Sheet Metal Forming,* Reinhold, New York.

Schey, J. A., 1983. *Tribology in Metalworking: Friction, Lubrication and Wear,* American Society for Metals, Metals Park, Ohio.

Springborn, R. K., ed., 1966. *Cold Bending and Forming Tube and Other Sections,* Society of Manufacturing Engineers, Dearborn, Michigan.

Tool and Manufacturing Engineers Handbook, 4th ed., Vol. 2, *Forming,* 1984. Society of Manufacturing Engineers, Dearborn, Michigan.

Wusatowski, Z., 1969. *Fundamentals of Rolling,* Pergamon, Oxford.

3

The Mechanics of Lubrication in Metal Cutting

Material removal processes are among the most important manufacturing operations. These processes are generally divided into two categories: metal cutting (chip generation) processes involving cutting on a lathe, milling, planing, shaping, broaching, reaming, sawing, drilling, grinding, and others, and alternative material removal processes utilizing electrical, chemical, optical (laser), and thermal techniques.

Lubrication is an important aspect of cutting and grinding processes, since a tool is in contact with the workpiece, and the interface is subjected generally to the same phenomena as it is in metal forming operations, described in Chap. 2. One major difference is that, in cutting, the tool generates new surfaces and hence the tool-chip and tool-workpiece interfaces are more critical. New surfaces do not have the protective surface layers, such as oxides, lubricants, and contaminants, that are generally present on the workpiece in forming operations.

Thus, unless lubricated properly, the interface between the cutting tool and the new surfaces generated is an area of high friction, elevated temperature, and, consequently, wear. Typically in cutting, all these properties are higher than those that occur during a forming operation. The rate of deformation in metal removal is generally higher than the rates of deformation in metal forming. Hence, lubrication mechanisms that rely on surface reactions may become less effective because of reduced reaction time in cutting operations.

In this chapter, the mechanics of lubrication in metal cutting and grinding processes is discussed in terms of chip formation, forces and energy requirements, temperature, surface finish and integrity, and dimensional accuracy of the machined parts.

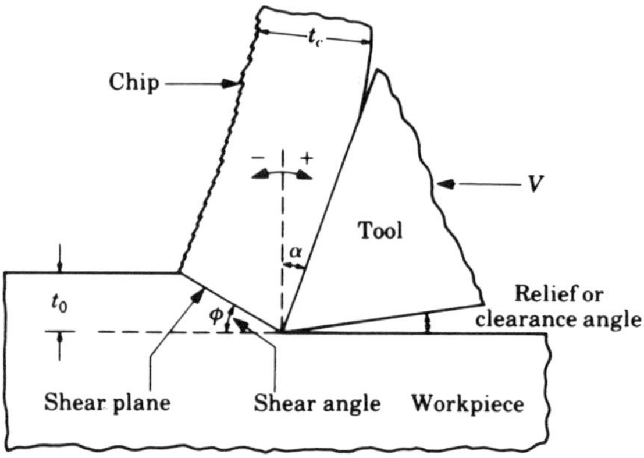

Figure 3.1 Schematic illustration of the cutting process with a single-point tool. (S. Kalpakjian, 1984. *Manufacturing Processes for Engineering Materials*, Addison-Wesley, Reading, Massachusetts, p. 463.)

3.1 MECHANICS OF CHIP REMOVAL

When a cutting operation is investigated at close range it is found that material removal takes place by a chip generation process, which can be schematically represented by Fig. 3.1. A tool of specified material, geometry, surface finish, and tip radius moves across a layer of material of depth of cut t_0, at a velocity (cutting speed) of V.

A chip is generated by a shearing mechanism, similar to a deck of cards or blocks sliding along each other (Fig. 3.2). By virtue of this mechanism, it can be seen that the thickness of the chip is greater than the depth of cut. The plane across which the chip is generated is called a shear plane and makes an angle ϕ with the direction of cut. The face of the tool (rake face) makes an angle α with a vertical line. This angle is called the rake angle and is one of the important variables in the cutting process.

As the rake angle increases, the shear angle increases and hence the chip becomes thinner. On the other hand, the rake angle can be set at a negative value, as a result of which the chip becomes thicker. Such a negative rake angle may be desirable in machining high-strength materials to prevent the tool from chipping due to the small angle of the tool tip.

After the chip is generated at the shear plane, it moves up along the face of the tool for a certain distance and then curls away. There is friction at the tool-chip interface due to the pressure and motion of the chip along the rake face. This friction contributes significantly to forces, temperature rise, and

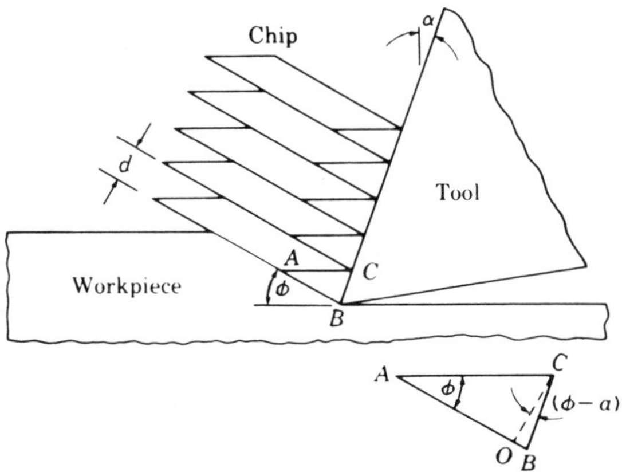

Figure 3.2 Mechanics of chip formation by cutting with a tool. (S. Kalpakjian, 1984. *Manufacturing Processes for Engineering Materials,* Addison-Wesley, Reading, Massachusetts, p. 464.)

the shape of the chip. If friction is high, the tool side of the chip can undergo further plastic deformation in the so-called secondary shear zone; the shear plane is the primary shear zone.

3.1.1 Forces and Energy Requirements

The forces involved in the deformation zone in cutting are shown in Fig. 3.3. The tool forces involve a cutting force (in the direction of tool movement) and a thrust force (perpendicular to the machined surface). The cutting force supplies the energy required to cutting. The thrust force does not contribute to this energy. It is, however, an important force component, as it tends to push the tool away from the surface. The greater the thrust force, the more rigid the machine tool should be to maintain dimensional accuracy.

There are two regions in cutting that dissipate energy: one is the shear zone (primary), where plastic work of deformation is dissipated; the other is the tool-chip interface (secondary), where frictional work is dissipated. Furthermore, if the tool tip is dull (large radius), or if it is worn, additional forces are required, thus increasing energy consumption.

It is apparent that if friction at the tool-chip interface is reduced, energy consumption will be reduced. Although it may not be recognized at first, friction also has an effect on energy consumption in the primary shear zone. This is because the shear angle is, among other factors, also influenced by friction.

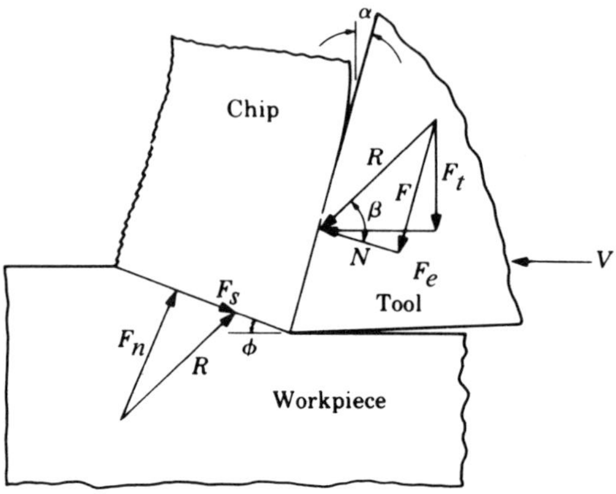

Figure 3.3 Schematic illustration of forces in cutting. (S. Kalpakjian, 1984. *Manufacturing Processes for Engineering Materials,* Addison-Wesley, Reading, Massachusetts, p. 470.)

As it can be intuitively seen in Fig. 3.3, the higher the friction, the more difficult it is for the chip to move up the face of the tool, and the chip becomes thicker. Consequently, the shear angle is also lower. Thus, an effective cutting fluid that reduces friction produces thinner chips, with lower force and energy requirements.

The work of plastic deformation in the primary shear zone is a direct function of the shear strain in this zone. The lower the shear angle (the thicker the chip), the higher is the shear strain, and hence the higher is the energy required. It can now be seen that by reducing friction at the tool-chip interface, we decrease the shear strain and thus reduce energy consumption. This reduction is, of course, in addition to the reduction in the work required to overcome friction at the tool-chip interface.

Unlike many metal forming processes, the estimation of forces and energies involved in cutting operations present considerable difficulties. Information in this regard is available in the literature based on experimental data. These data can also be useful in estimating the maximum rate of material removal possible on a particular machine tool.

3.1.2 Effects of Friction on Chip Morphology

In addition to its effects on forces and energies involved, friction also affects chip morphology, that is, the shape of the chip. Depending on tool geometry,

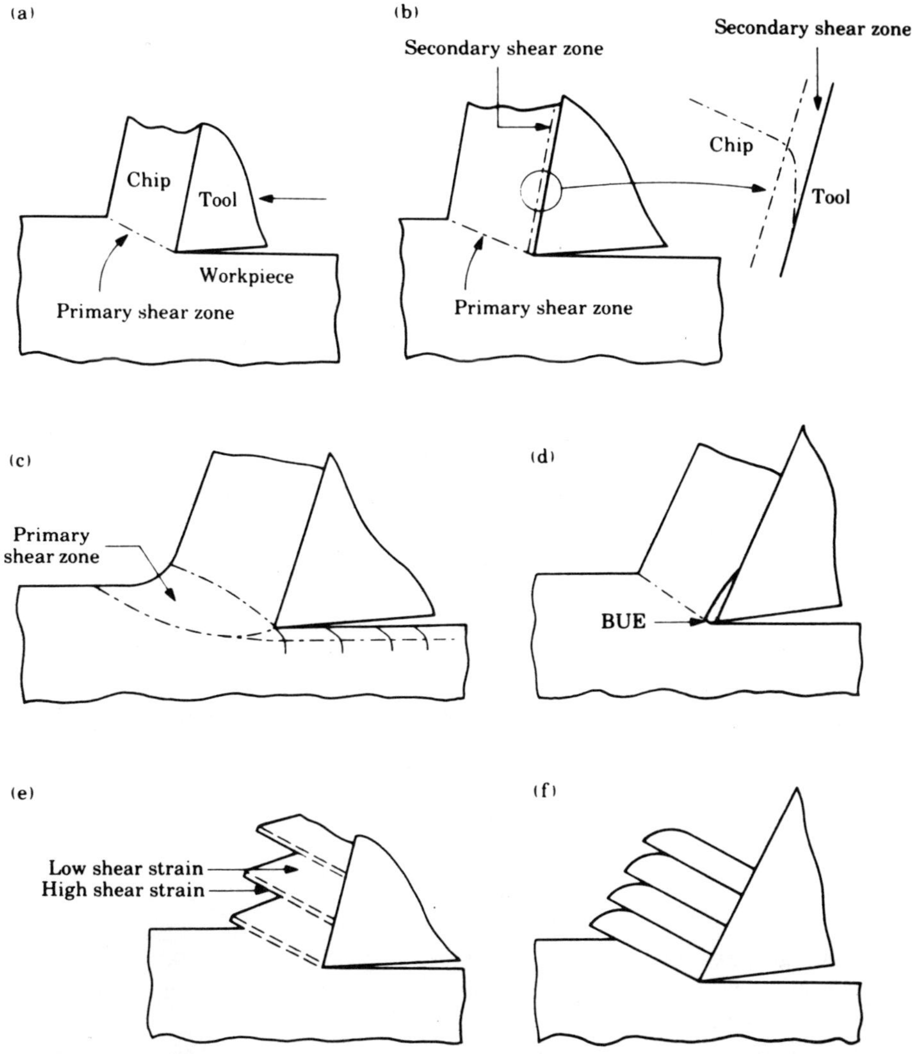

Figure 3.4 Types of chips obtained in metal cutting. (a) Continuous chip. (b) Continuous chip with a secondary shear zone. (c) Continuous chip with a large primary shear zone. (d) Built-up edge in a continuous chip. (e) Inhomogeneous (serrated) continuous chip with regions of low and high shear in the primary zone. (f) Discontinuous chip. (After M. C. Shaw. From S. Kalpakjian, 1984. *Manufacturing Processes for Engineering Materials,* Addison-Wesley, Reading, Massachusetts, p. 467.)

process parameters, and the properties of the workpiece material, various types of chips can be produced in cutting. The basic types of chips are continuous, discontinuous, segmented, and the development of a built-up edge (BUE) on chips (Fig. 3.4). It is beyond the scope of this book to describe various details of chip morphology, except to emphasize the role of friction and lubrication.

Continuous chips are desirable because the cutting process is at a steady state. In other words, forces, chip morphology, and surface finish do not change as the cutting operation progresses. (This is not the case with other types of chips, as described later.) Lubrication contributes to the formation of this type of chip by reducing friction and increasing the shear angle.

Discontinuous chips are generally due to insufficient ductility of the workpiece material and .o process variables, such as depth of cut and cutting speed. Lack of ductility means that the cutting process cannot take place in a steady state because the material is unable to undergo the large shear strains required in generating continuous chips. Process variables affect this type of chip by indirect effects through temperature and the stresses in the cutting zone. As a practical consideration, discontinuous chips are preferred because of ease of removal and lack of interference with mechanical components of the machine tool.

The built-up edge on the tool is significant because of its adverse effect on surface finish and integrity. A BUE consists of material from the workpiece that has adhered to the tool tip, thus changing its geometry and affecting the mechanics of chip formation. Generally, a BUE is unstable; that is, its size and shape vary during the cut. During the cutting process, parts of the BUE are deposited on the machined surface, thus contributing to surface roughness. The BUE is influenced by many factors; among them is lubrication. An effective cutting fluid can reduce or eliminate a BUE.

3.1.3 Temperature

Friction, either directly or indirectly, also increases the temperature during cutting. This increase is because most of the energy dissipated in the operation is converted into heat.

It has been shown experimentally that temperature rise is most severe at about mid point at the tool-chip interface. It can reach 2000°F (1100°C), indicating also the importance of the selection of proper tool materials and the role of cutting fluids as coolants.

Temperature in cutting is an important parameter, as it affects tool life, dimensional accuracy, and tolerances of the part being machined. High temperatures on the rake face cause crater wear; thus, the hot hardness of the tool and its reacitvity with the workpiece material are important contributing factors. As discussed in Sec. 3.14, the importance of cooling is thus self-evident.

Although crater wear generally occurs at higher cutting speeds, flank wear (at the bottom surface of the tool rubbing along the machined surface) is always present in cutting. This type of wear is most influenced by cutting speed and less so by process parameters, such as feed and depth of cut. Cutting fluids also have an effect in reducing flank wear by lowering the temperature in the cutting zone and by reducing friction at the tool-chip interface.

A portion of the heat generated during cutting goes into the workpiece and causes its expansion. After the workpiece cools, the machined dimension may be smaller than the desired diameter if the thermal effect is not compensated for. Excessive temperature on the machined surface may also cause phase transformations in the material. (This effect is particularly significant in grinding, as described in Sec. 3.2.)

3.1.4 Role of Cutting Fluids

It is apparent from the foregoing discussion that cutting fluids, by virtue of their lubricating and cooling capabilities, can effectively influence the mechanics of cutting. Cutting fluids improve tool life by reducing friction, wear, and temperature. They also reduce forces and energy consumption and improve surface finish and integrity.

Friction and wear are reduced by virtue of both the lubricating and cooling effects of the cutting fluid. Cutting forces and energy requirements are reduced because friction is reduced, thus requiring less energy. Surface finish is improved primarily because of the elimination, reduction, or increased stability of the built-up edge, which is significantly dependent upon the effectiveness of the cutting fluid. The lubricant reduces friction and, hence, the tendency for adhesion at the tool-chip interface. Furthermore, it cools the cutting zone.

Vibration and chatter in cutting processes are also reduced or eliminated by lubrication because of the reduction in forces and friction. As a consequence of the foregoing effects, surface integrity is also improved.

The cooling capacity of cutting fliuds reduces the temperature and distortion of the parts, even though generally a small portion of the heat generated in the cutting process goes into the workpiece, especially at high cutting speeds. (Most of the heat is carried away by the chip; thus, the chip is a very effective heat sink in cutting.) This is important in controlling the tolerances and dimensional accuracy of machined parts. Cutting fluids also help wash away chips from the cutting area and protect the newly machined surfaces from corrosion and environmental contamination. The effectiveness of a cutting fluid depends on the method of application (Chap. 8), cutting speed, type of machining operation, and the temperatures encountered in the operation.

If the cutting fluid is oil based, it is primarily a lubricant. On the other hand, water-based cutting fluids are not only effective lubricants but also excellent coolants by virtue of the high specific heat and thermal conductivity of water. It has been shown that the temperature in cutting increases substantially as cutting speed increases. Thus, at high cutting speeds, generally above 100 ft/min (0.5 m/sec), cooling the tool-chip interface is of primary importance. At low speeds lubrication may be more important than cooling, since the temperature rise is not significant.

The manner in which the fluid enters the cutting zone and influences the mechanics of chip generation has been studied extensively. Experiments have indicated that the fluid enters the tool-chip interface from the sides of this zone, rather than in a direction against the flow of the chip up the rake face of the tool. This penetration requires that the molecular size and the surface tension characteristics of the fluid be controlled properly. It has been established that the lubricating properties of fluids in cutting arise from the formation of films of low shear strength and that chemical reactions may take place due to the high reactivity of the freshly machined surfaces (Chap. 5).

3.2 ABRASIVE PROCESSES

Abrasive processes are another group of important material removal operations. Although they are generally regarded as finishing operations, abrasive machining also involves removal of material in large volumes. In contrast to the cutting operations just described, abrasive removal processes involve abrasive grains of high hardness that are small in size, with irregular shapes. Commonly used abrasives are aluminum oxide, silicon carbide, cubic boron nitride, and diamond. Another major difference from other cutting processes is that grinding speeds are generally much higher than the speeds employed in cutting operations.

Although the shape of abrasive grains is difficult to define geometrically, observations of abrasive processes have indicated that the mechanics of removal is similar to cutting and that it can be represented schematically by Fig. 3.5. The grains are generally very small and the depth of cut even smaller, of the order of 0.001 in. (0.025 mm). Because of their shape, grains generally have a highly negative rake angle; thus the shear angle is smaller

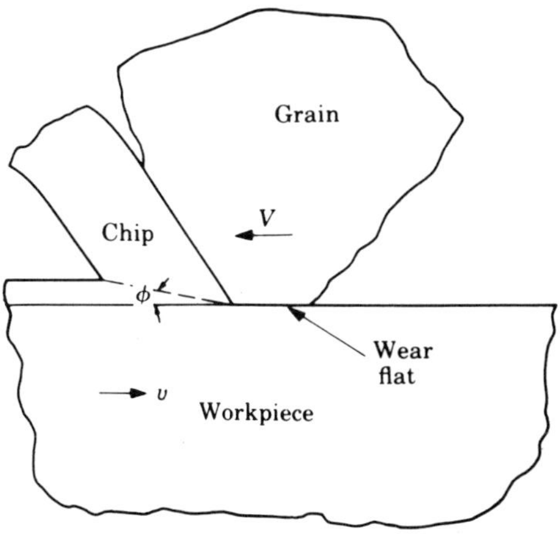

Figure 3.5 Abrasive grain and wear flat in grinding. (S. Kalpakjian, 1984. *Manufacturing Processes for Engineering Materials,* Addison-Wesley, Reading, Massachusetts, p. 565.)

than in cutting with a tool. This also means that the work of plastic deformation is higher in generating a grinding chip.

A chip in grinding is produced in a manner similar to that in cutting. Because of the rough outline of the grain, friction at the grain-chip interface can be significant, again indicating the importance of lubrication at this interface. Unlike in cutting, it is important to have sufficient space (open structure) on the surface of a grinding wheel to accomodate the chips being generated. Otherwise, the wheel will become loaded.

An important aspect in the abrasive removal mechanism is the wear flat that develops on the grains, similar to the flank wear of cutting tools. This wear is known as attritious wear and is due to complex physicochemical reactions between the workpiece and the grain, with temperature having a significant effect on this reaction. As can be seen, a wear flat rubs along the surface being generated (Fig. 3.5), thus involving friction where the frictional work is converted into heat. This heat source can cause a severe rise in surface temperature of the workpiece and, unless controlled properly, can cause thermal cracks, burns, residual stresses, and phase transformations on the ground surface. The importance of the cooling capabilities of grinding fluids thus becomes apparent.

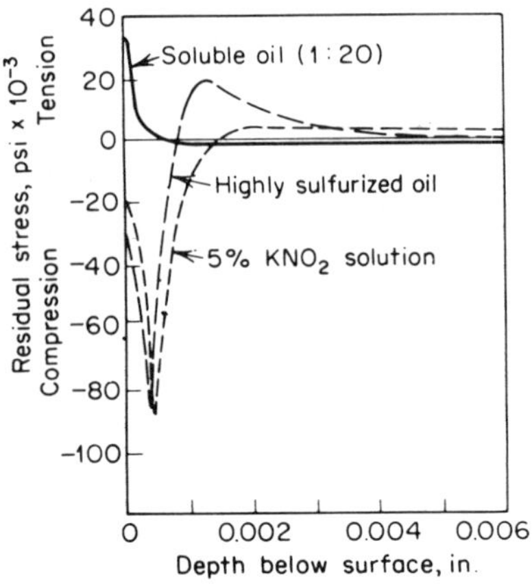

Figure 3.6 The effect of grinding fluid on residual stresses for tungsten. (After N. Zlatin, M. Field and J. V. Gould)

A significant effect of wear flats (and the overall grinding process) is the generation of residual stresses on the surface of ground parts. These stresses are caused by surface plastic deformation and by temperature variations during grinding. Tensile residual stresses are detrimental to the fatigue life of ground components. Although also important in cutting operations, residual stresses are even more important in grinding since ground parts are generally subjected to more critical applications involving cyclic loading and stresses.

Residual stresses are affected by the type of grinding fluid, the type of grinding wheel (especially bond strength), and such operating parameters as speed and depth of cut. It has been shown that, by changing the type of grinding fluid employed, the nature and distribution of residual stresses can be completely reversed (Fig. 3.6). These effects are due to the different cooling and lubricating capabilities of the fluids employed and the mechanisms by which they reduce friction and attendant plastic deformation. Techniques have been developed, known as low-stress or gentle grinding, to reduce detrimental residual stresses in abrasive processes. This operation generally involves using softer wheels, lower wheel speeds, and higher work speeds.

A third significant aspect of temperature rise in abrasive processes is its effect on part expansion and distortion. Because most grinding operations are for finishing purposes, often requiring extreme accuracy, thermal expansion must be avoided using grinding fluids with the proper combination of cooling capacity and lubrication. It should be pointed out that temperature rise can also be reduced by varying the following process parameters: reducing the depth of cut and wheel speed, increasing the work speed, reducing wheel hardness, and controlling the wheel composition.

3.2.1 Effects of Grinding Fluids

Grinding fluids have functions that are similar to those for metal cutting, as in turning, milling, and drilling. Compared with cutting operations, grinding involves higher temperatures and speeds and smaller dimensions of cut. Furthermore, since grinding is an important finishing and precision operation, part distortion and surface finish and integrity are generally more critical than in cutting.

Grinding fluids prevent temperature rise not only by reducing the cutting energy by lowering friction but also by cooling the workpiece. They also prevent loading the wheels, that is, filling the pores of the wheel with grinding debris. If the wheel is loaded, friction and grinding forces increase, temperature increases, surface burn may occur, and the workpiece surface may undergo phase transformations, as well as develop high tensile residual stresses. Loaded wheels also reduce grinding efficiency by altering the surface texture of the wheel. Thus, effective grinding fluids keep the wheel surface clean. This requires specially designed nozzles.

The lubricating capabilities of grinding fluids are also important in reducing friction. This, in turn, means lower forces (hence, lower deflection of the parts ground, especially in precision grinding), lower energy consumption, and lower temperatures. It has been shown experimentally, for example, that the temperature rise in grinding with a compounded oil may be less than when using an oil-in-water emulsion. Thus, even though the latter has a higher cooling capacity, the reduction in friction by the oil (hence less energy consumption and less heat generated) has a greater overall effect.

The exact mechanisms by which grinding fluids affect the cutting operation are complex and not fully understood. They depend on the type of abrasive grain, the bond, the type of workpiece and its reactivity with the abrasive grain at elevated temperatures (since instantaneous temperatures at the grit-chip interface can be very high), and such operating parameters as speed and feed. The attritious wear of abrasive grits, for instance, is both physical and chemical in nature. It should also be noted that although an individual grit is in contact with the metal surface only during a very small portion of its cycle (one revolution), the grit is in constant contact with the grinding fluid. This allows chemical reactions as well as adsorption of the fluid to take place on the grit surfaces.

Each type of workpiece material requires a particular type of abrasive, bond, and various other wheel specifications, such as structure, as well as a particular grinding fluid.

BIBLIOGRAPHY

Armarego, E. J. A., and R. H. Brown, 1969. *The Machining of Metals,* Prentice-Hall, Englewood Cliffs, New Jersey.

Bartz, W. J., and J. Wolff, eds., 1982. *Lubrication in Metal Working,* 2 vols., Technische Akademie Esslingen, Germany.

Boothroyd, G., 1975. *Fundamentals of Metal Machining and Machine Tools,* McGraw-Hill, New York.

Cutting and Grinding Fluids, 1967. Society of Manufacturing Engineers, Dearborn, Michigan.

Kalpakjian, S., 1984. *Manufacturing Processes for Engineering Materials,* Addison-Wesley, Reading, Massachusetts.

Machinery's Handbook, 21st ed., 1980. Industrial Press, New York.

Machining Data Handbook, 3d ed., 2 vols., 1980. Machinability Data Center, Cincinnati, Ohio.

Metals Handbook, 8th ed., vol. 3, *Machining,* 1967. American Society for Metals, Metals Park, Ohio.

Schey, J. A., 1983. *Tribology in Metalworking: Friction, Lubrication and Wear,* American Society for Metals, Metals Park, Ohio.

Shaw, M. C., 1957. *Metal Cutting Principles,* 3d ed., MIT Press, Cambridge, Massachusetts.

Tool and Manufacturing Engineers Handbook, 4th ed., vol. 1, *Machining,* 1983. Society of Manufacturing Engineers, Dearborn, Michigan.

Trent, E. M., 2nd ed., 1984. *Metal Cutting,* Butterworth, Woburn, Massachusetts.

Zintak, D., ed., 1982. *Improving Production with Coolants and Lubricants,* Society of Manufacturing Engineers, Dearborn, Michigan.

Zorev, N. N., 1966. *Metal Cutting Mechanics,* Pergamon, Oxford.

4

Surfaces and Their Relationship to Lubrication

As outlined in Chaps. 2 and 3, friction can have a significant effect on the mechanics of metal forming and machining processes. Since friction occurs between rubbing surfaces (tool and workpiece), it is essential to evaluate surfaces as they influence friction, wear, and lubrication. A surface, in this context, is not only the surface as it is visible to the naked eye, but it must be considered to consist of some thickness of material, since the subsurface influences the behavior of the surface layer as well. The thickness of this layer, which is critical for metalworking operations, depends on the particular process by which the surface is generated or loaded.

The properties that are important are physical, chemical, metallurgical, and mechanical. Which of these properties is singificant with respect to the surface of a die or workpiece depends not only on the particular metalworking process but also on the range of process parameters that are operational, such as speed, temperature, time, and degree of plastic deformation.

4.1 SURFACE CHARACTERISTICS

Although a surface that is freshly generated (by an ideal process that does not alter its characteristics) may be regarded as having the same properties as the bulk material, over a period of time and depending on the environment the same surface may acquire very different properties. Furthermore, in actual processing, forming or removal of material is never ideal and the operation alters the surface properties as a result of plastic deformation and temperature. To fully appreciate these effects, let us inspect Fig. 4.1.

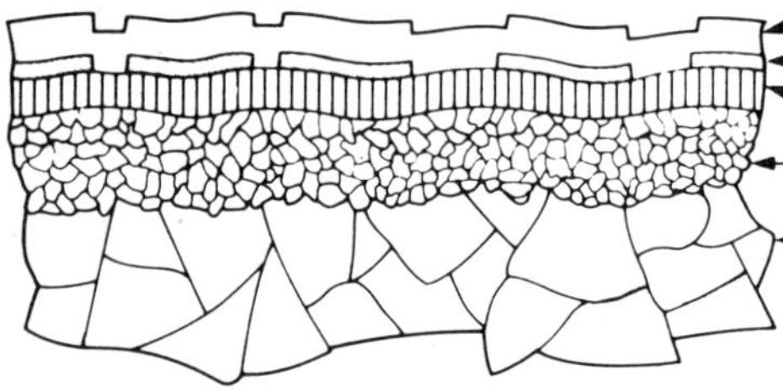

Figure 4.1 Schematic illustration of various layers on the surface of a metal. (After E. Rabinowicz. From S. Kalpakjian, 1984. *Manufacturing Processes for Engineering Materials,* Addison-Wesley, Reading, Massachusetts, p. 178.)

A surface is composed of various layers, starting with the bulk metal. Above this bulk is a layer of material that is work hardened due to plastic deformation during processing (also referred to as the Beilby layer). The depth of this layer depends, for instance, on the friction at the interface between the die and the workpiece, or, if machined, on the sharpness of the tool and the control of various other process parameters during machining and the use of cutting fluids. This layer includes defects on the atomic scale, such as dislocations, vacancies, and voids, as well as residual stresses and possibly phase transformations. On a macroscale, layers of material may have melted and resolidified, and surface cracks may have developed during thermal cycling of the workpiece.

Above the work-hardened layer is an oxide layer, which in almost all cases, except when the metalworking process is carried out in vacuum, is inevitable. The thickness and chemistry of this layer depend on the material and its environment. They may also depend on the hardness, structure, and surface finish of the work-hardened layer.

This oxide layer may be covered with adsorbed layers or films of various contaminants, gas, and moisture – their nature depending on the particular environment. These contaminants include any matter deposited on the surface, such as dirt, grease, rust preventatives, cleaning compound residues, and paint. Furthermore, various elements or particles may also be deposited on the part from the environment, such as pollution from the atmosphere (SO_3, NO_2, or other compounds) or from inside a plant or manufacturing facility.

In addition to their physical and mechanical characteristics, surfaces are also geometric entities with certain features that are important to lubrication (see Sec. 4.2). It is also evident that the proper use and application of a lubricant to surfaces requires careful consideration of these characteristics and the interactions that they may undergo with the lubricant. In the sections that follow some details of these surface characteristics are described as they affect lubrication in metalworking.

4.1.1 Surface Chemistry

The surface of a metal is comprised of metal atoms bounded by air (nitrogen, oxygen, and carbon dioxide) and other nonmetallic atoms. The chemistry of the surface may often involve a gradient in its composition. Thus, for instance, the carbon concentration in steel may be higher or lower at its surface than at its bulk. This gradient will depend on whether the surface has undergone decarburization during processing or carburization during subsequent heat treatment.

Because of reactions with air, the metals commonly used in manufacturing processes have surfaces that are not purely metallic in nature. Surfaces are generally a mixture of oxides and hydroxides. Depending on the environmental conditions, chloride and sulfide compounds may also form on the surfaces. Iron, for instance, has a complex oxide structure with FeO adjacent to the iron, followed by a layer of Fe_3O_4, and then a layer of Fe_2O_3 exposed to the environment. Furthermore, the carbon in steels is preferentially oxidized, thus leaving a carbon-depleted layer on the surface adjacent to the FeO layer.

In nonferrous metals, aluminum in air has a dense amorphous layer of Al_2O_3 (10 to 20 Å thick), with a thick, porous hydrated region adjacent to the air. Copper in air has a Cu_2O layer adjacent to the copper surface, followed by a layer of CuO exposed to the air. Various other metals form oxides as a result of their reaction with oxygen and water vapor.

The rate of oxidation increases with increasing temperature and the partial pressure of oxygen. Although slowly grown oxide films are smooth and adhere strongly to the base metal, oxide films formed at higher temperatures have a rough profile. A thick oxide film tends to develop cracks and lift itself from the surface of the base metal in the form of blisters.

It should be noted that oxides may be regarded either as contaminants or as protective films acting as a barrier to the reaction of the base metal with oxygen. Oxides are generally hard and brittle; hence, their abrasive behavior at die-workpiece interfaces must be considered accordingly. As we shall see in Chap. 5, oxides may also play a significant beneficial role in friction and wear. The presence of oxide films also affects the bond strength of the lubricant and its behavior on the surface of the workpiece.

During further processing of a metal covered with an oxide layer, the oxide may be embedded beneath the surface, thus affecting the surface quality of the product. It is also possible that the oxide layer becomes detached and fragments, thus producing a wear particle.

Furthermore, because of the nature of oxide films, an oxidized metal surface may be less likely to adhere (weld) to tools and dies, thus reducing friction. However, since oxides are generally harder than the base metals, wear may increase at interfaces where abrasive oxide layers are present. In lubricated interfaces, the inherent roughness of oxide layers serves as reservoirs and traps lubricants, just like the surface roughness produced from a machining operation.

4.1.2 Surface Structure

Plastically Deformed Layer (Texture)

The surface of the die (or tool) and the workpiece in metalworking processes is subjected to relative motion, generally under variable pressure and a temperature gradient. The geometry of the die-workpiece interface in forming and tool-workpiece interface in machining operations is often complex, leading to nonuniform deformation of surfaces in the immediate vicinity of the deformation zone. For instance, if friction at a die-workpiece interface is high, the workpiece surfaces will be deformed severely, generating a work-hardened layer. If the temperature rise due to high friction is high, the surface may undergo metallurgical changes, such as phase transformation, and even burning and melting. The surface may also undergo a burnishing action. In a process such as grinding, if the wheel becomes dull (Fig. 3.5) by having the abrasive grits develop wear flats that rub against the surface being generated, then the surface will be subjected to severe shear strains.

Plastic deformation of the surface results in a texture different from that of the bulk of the workpiece. Furthermore, an orientation of the surface structure will result from an overall deformation, which is generally different from the original orientation. Depending on the severity of the deformation, dislocations and vacancies are generated along a surface layer with thickness a function of the degree of plastic deformation. Thus, along with the chemistry gradient, one can expect a gradient of a defect structure as well as nonuniformity of such defects to some depth below the surface. These changes, in turn, have an effect on the various properties of the surface, such as its reactivity and friction and wear characteristics. The hardness, orientation, and surface finish of both tool and workpiece directly influence lubricant wetting, reactions, and film thickness characteristics. These effects must be taken into account in the proper selection of lubricants.

Residual Stresses

One effect of deformation during metalworking is to develop residual stresses on the surface of the workpiece. Residual stresses are important because of their effect on fatigue strength and in the control of flatness and size. Tensile residual stresses are detrimental to fatigue life; also, they may reduce tool and die life. On the other hand, compressive residual stresses on a surface are beneficial in improving fatigue strength.

Residual stresses generally have two major sources. One is nonuniform deformation, such as by working the surface, as in machining with a dull cutting tool (see Chap. 3). A dull tool tends to rub over the surface and produce severe shear strains. However, in most metalworking processes, especially in bulk deformation, such as in extrusion and rod and wire drawing, the material is generally subjected to large nonuniform deformation. This is due to the deformation geometry in the die region and to friction, and it cannot be avoided. The nature of residual stresses (whether tensile or compressive) and their magnitude and distribution also depend on process parameters, such as reduction per pass.

A second source of residual stresses is the development of a thermal gradient due to a nonuniform temperature distribution within the bulk of the workpiece. In a casting, for instance, the outer surfaces cool first, thus shrinking first, while the inner bulk is still hot. Subsequently, the inside begins to cool and, while shrinking, further shrinks the already solidified outer skin. This thermal cycling results in a casting that, when completely cold, develops compressive residual stresses on its outside and tensile stresses on the inside. In metalworking processes in which the temperatures generated are significant, residual stresses are generally due to both deformation and temperature gradients.

Residual stresses can be significant in lubrication because the reactivity of chemicals and various chemical compounds can be influenced by these stresses. Thus, for instance, it has been shown that tensile residual stresses on a surface increase reactivity. Surfaces oxidize more rapidly if they are subjected to tensile residual stresses. The energy for these processes is derived from the strain energy of a stressed volume of material at the surface. Because a hot-worked surface generally has a lower level of residual stresses compared with a cold-worked surface, a hot-worked surface is relatively less reactive.

4.2 SURFACE FINISH

Since lubrication involves interfacial phenomena, the surface finish of both the workpiece and the tools and dies is an important aspect of metalworking

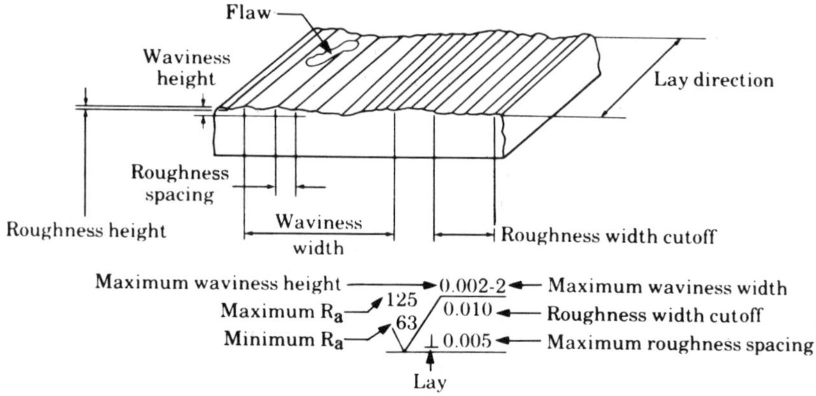

Figure 4.2 Surface finish terminology and symbols. (S. Kalpakjian, 1984. *Manufacturing Processes for Engineering Materials,* Addison-Wesley, Reading, Massachusetts, p. 181.)

lubrication. Surface finish may be measured or indicated in terms of roughness, waviness, lay, and flaws (Fig. 4.2).

The roughness measurements are given either as an arithmetical average R_a (AA; also known as CLA, center-line average), or root-mean-square (RMS) average, identified as R_q. The measurement and interpretation of actual surface roughness are difficult, primarily because two surfaces may have the same roughness value, yet their topography may be quite different (Fig. 4.3). Furthermore, surface roughness profiles as obtained from profilometers depend on the stylus tip radius and its geometry.

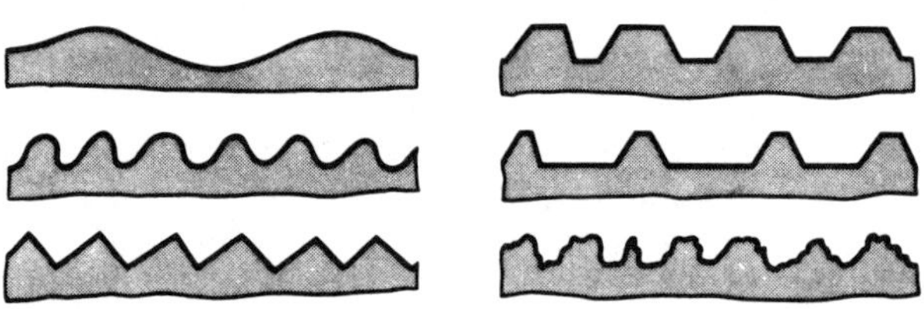

Figure 4.3 Various surface profiles with the same surface roughness value. (S. Kalpakjian, 1984. *Manufacturing Processes for Engineering Materials,* Addison-Wesley, Reading, Massachusetts, p. 183.)

Although a rough surface can have an adverse effect on mechanical properties, especially fatigue strength, it can provide a reservoir of lubricant in a metalworking operation. The lubricant can be trapped in small surface pockets and also contribute to supporting the load at the tool-workpiece interface while supplying lubricant during the deformation process.

Depending on operating conditions and the materials involved, an originally smooth surface may become rough during metal forming processes. The roughness produced depends on the strain that the workpiece surface has undergone, its crystal structure and grain size, the properties of the lubricant, such as its viscosity, and the formation of lubricant pockets at the interface. On the other hand, an initially rough workpiece surface may become smoother under high contact stresses with the dies, particularly in the case of boundary or mixed-film lubrication (Chap. 5).

The direction of roughness (lay) is also an important factor. Generally, a roughness direction perpendicular to the relative sliding direction between the die and the workpiece is desirable for trapping lubricants. A surface finish parallel to the sliding direction is not as effective.

Comparatively, it is generally desirable for the workpiece surface to be rougher than the tools or dies as a means of retaining lubricants at the interface. Otherwise, the workpiece surface may be damaged by the harder rough die, particularly if any region of the interface is starved for lubricant and if there are asperities on the die surface. On the other hand, a moderately rough die surface may provide reservoirs for retaining lubricants, particularly semisolid or solid lubricants, such as graphite. (Some of these aspects are discussed in Chap. 5, lubrication fundamentals as influenced by surface roughness, Sec. 5.3).

4.3 SURFACE PREPARATION

Workpieces for forming and machining operations have surface characteristics that are a result of prior metalworking processes. For example, sheet metal may be used in the as-rolled condition. Metal cutting operations may be performed on surfaces generated from a previous process, such as forging, extrusion, or drawing. The effectiveness of a lubricant depends on the nature of these surfaces (physical, chemical, or mechanical).

Prior to a particular forming or machining operation, workpiece surfaces may be conditioned by various other operations. Some of these are as follows.

A cylindrical workpiece, for instance, that has been produced by a process such as hot extrusion may not have the necessary tolerances or the surface texture suitable for a cylindrical grinding operation. In such a case, the appropriate surfaces of the workpiece are first machined by turning in preparation for subsequent grinding. Likewise, in a heading operation on a bar, the ends of the bar may not be sufficiently square if cropped by shears.

An important aspect of machining for such applications is the removal of oxides from a surface. Oxides may be abrasive, thus causing excessive wear of tools. They may also interfere with the role of surface-active lubricants, which are most effective on clean surfaces.

In shot blasting, heavy oxide layers and other physical contaminants are removed from a surface by the mechanical impact of shots. Because of the mechanisms involved, shot blasting leaves a surface with characteristics different from those of a machined surface. This, in turn, affects the performance of the lubricants, both in their chemical activity with the surface and in lubricant entrapment. The presence of shot embedded on the workpiece should also be considered.

Another method is sandblasting, in which surfaces are cleaned by the abrasive action of sand. The effect on surface finish and characteristics depends on the velocity and the angle at which sand is directed on to the surface. Compared with shot blasting, which involves larger masses of particles, sandblasting involves much lower inertia and less plastic deformation of the surface layers of the workpiece.

The purpose of pickling is to chemically remove oxides and other contaminants from a surface. This is usually done with a dilute solution (10 to 25%) of sulfuric, hydrochloric, nitric, phosphoric, or hydrofluoric acid. Because pickling is a chemical treatment, the effectiveness of a lubricant applied to a pickled surface depends on its cleanliness, whether or how well it has been rinsed, the nature of any residues left on the surface, and how much, if any, etching of the surface has occurred during pickling.

In addition to the foregoing methods of surface preparation, other techniques are also available, such as various chemical cleaning solutions and cleaning with flame, steam, ultrasonics, and vapor degreasing. All these methods produce distinctive surface characteristics on the workpiece.

4.4 SURFACE COATINGS AND TREATMENTS

Prior to various metalworking operations, parts may have been

1. Painted (such as for sheet-metal containers)
2. Coated with polymers and other nonmetallic materials (such as laminated sheet metal for decorative or functional purposes)

3. Plated with metal, either electrolytically or by other means, such as hot dipping (tinplate, galvanized steel, and chrome- or nickel-plated parts)
4. Heat treated by carburizing, nitriding, or other processes (such as in gears, cams, shafts, and bearing races)

The first three examples generally involve subsequent forming operations in which the pressures and temperatures at the tool, die, and workpiece interfaces are not particularly high. Nevertheless, effective lubrication is important in reducing surface damage to the parts. The lubricant interaction in these operations depends greatly on the chemical and mechanical nature of the coating and its physical characteristics. If the coating is not reactive with the die material, or if it has inherent lubricity, as do some polymers (polytetrafluoroethylene; PTFE) and low-shear-strength metals (tin), friction and wear are likely to be low. Each situation should be analyzed individually. Some of the relevant details are covered in Chap. 5.

Heat-treated surfaces, on the other hand, are generally followed by grinding operations to produce the final shape of the part with proper tolerances. The brittle nature of heat-treated surfaces precludes any subsequent forming operations. The role of grinding fluids as lubricants in grinding heat-treated surfaces is essentially the same as that described in Chap. 3. Fluids should be effective coolants and applied properly to avoid grinding cracks and part distortion and to minimize residual stresses.

BIBLIOGRAPHY

Adamson, A. W., 1960. *Physical Chemistry of Surfaces,* Wiley Interscience, New York.

Appearance of Metallic Surfaces, STP 4, 1970. American Society for Testing and Materials, Philadelphia.

Bikerman, J. J., 1970. *Physical Surfaces,* Academic Press, New York.

Bowden, F. P., and D. Tabor, 1950, 1964. *The Friction and Lubrication of Solids,* 2 vols., Oxford University Press, London.

Braithwaite, E. R., 1964. *Solid Lubricants and Surfaces,* Macmillan, New York.

Buckley, D. H., 1981. *Surface Effects in Adhesion, Friction, Wear, and Lubrication,* Elsevier, New York.

Carbonara, R. S., and J. R. Cuthill, eds., 1976. *Surface Analysis Techniques for Metallurgical Applications,* STP 596, American Society for Testing and Material, Philadelphia.

Dowson, D., C. M. Taylor, M. Godet and D. Berthe, eds., 1978. *Surface Roughness Effects in Lubrication,* Mechanical Engineering Publications, London.

Gatos, H. C., ed., 1960. *The Surface Chemistry of Metals and Semiconductors,* Wiley, New York.

Halling, J., 1975. *Principles of Tribology,* Macmillan, New York.

Iliuc, I., 1980. *Tribology of Thin Layers,* Elsevier, New York.

Kane, P. F., and G. B. Larrabee, eds., 1971. *Characterization of Solid Surfaces,* Plenum, New York.

Mittal, K. L., ed., 1979. *Surface Contamination,* Plenum, New York.

Moore, D. F., 1975. *Principles and Applications of Tribology,* Pergamon, New York.

Peterson, M. B., and W. O. Winer, eds., 1980. *Wear Control Handbook,* American Society of Mechanical Engineers, New York.

Properties and Metrology of Surfaces, 1967-1968. Proceedings of the Institution for Mechanical Engineering, vol. 182, Part 3K, London.

Quinn, T. F. J., 1971. *The Application of Modern Physical Techniques to Tribology,* Butterworths, London.

Rabinowicz, E., 1965. *Friction and Wear of Materials,* Wiley, New York.

Surface Technology, 1973. Proceedings of the International Conference, Society of Manufacturing Engineers, Dearborn, Michigan.

Thomas, T. R., and M. King, 1980. *Surface Topography in Engineering,* British Hydromechanics Research Association, Bedford, England.

Williamson, J. B. P., 1978. *Topography of Solid Surfaces – An Interdisciplinary Approach to Friction and Wear,* NASA SP-181, pp. 85-142.

5

Metalworking Lubrication Fundamentals

In metalworking operations involving forming or machining, the interface between the tools, the dies, and the workpiece is subjected to a wide range of variables. Temperatures may range from ambient, such as in cold forming, to around 4000°F (2200°C), as in hot extrusion. Speeds may range from 10^{-1} to 10 ft/min (10^{-4} to 10^{-2} m/sec), such as in superplastic forming of aluminum and titanium alloys, to around 10^5 ft/min (100 m/sec) in such operations as explosive forming, high-speed machining, and small-diameter wire drawing. Contact pressures at the tool-workpiece interfaces may range from elastic stresses of the order of 1000 psi (7 MPa), such as in sheet-metal forming, to multiples of the yield stress of the material, as in cold extrusion, that is, 350,000 psi (2400 MPa).

To ensure effective lubrication, such operating parameters as speed, temperature, and pressure, which are typical of the metalworking operation, must be taken into account. In fact, the magnitude of these parameters varies widely from one process to another. Furthermore, variations in these parameters occur during a single cycle of the operation, as well as nonuniformly at the die-workpiece interfaces or within the cutting zone in machining processes.

In addition to the operating variables, other factors also influence the effectiveness of lubrication. Among these are the composition of tool and workpiece materials and their structure, surface finish, and chemistry (oxides, contaminants, and adsorbed layers of gases; Chap. 4), as well as physical properties, tool design, and state of stress in the workpiece being produced.

Small variations in any of these conditions can have a major effect on friction, wear, and the effectiveness of lubrication.

In this chapter the fundamentals of metalworking lubrication are discussed in terms of material and process variables, including the effects of the chemistry and properties of the lubricants used.

5.1 FRICTION

To assess the need for lubrication and the type of lubricant required in a particular metalworking operation, it is first necessary to understand the nature and mechanisms of friction. Friction may simply be defined as the resistance to relative motion between two bodies in contact. Friction is an energy-dissipating process, as a consequence of which the temperature at the interface rises and, if excessive, can cause surface damage and also influence the deformation mechanism in a metalworking process.

The earliest theories proposed suggested that friction was due to the interlocking of the two rough surfaces sliding along each other. However, such an explanation cannot account fully for the observed dependence of friction on a large number of variables, such as load, speed, temperature, the materials involved in the sliding pair, and the various effects of fluids and gases at the interface. Subsequently, many models have been proposed to explain the nature of friction, with varying degrees of success.

The most commonly accepted theory of friction is based on adhesion between the asperities of the two contacting bodies. It has been shown by experimental techniques (such as electrical resistance measurement and optical methods) that, regardless of how smooth two surfaces are, they contact each other at only a fraction of their apparent area of contact (Fig. 5.1).

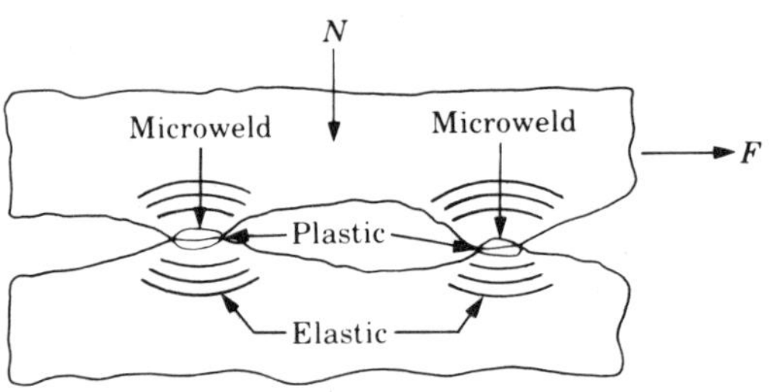

Figure 5.1 Schematic illustration of the interface between two surfaces in contact. (S. Kalpakjian, 1984. *Manufacturing Processes for Engineering Materials,* Addison-Wesley, Reading, Massachusetts, p. 196.)

Thus, the load is supported by relatively few asperities in contact; hence, the normal stress at these asperity junctions is high. Under light loads, the contact stresses may only be elastic. However, as the load increases (to the level of interest in metalworking processes), plastic deformation of the asperities takes place and the junctions form an adhesive bond (microwelds).

The nature and strength of this bond depend on many factors. Among these are the mutual solubility and diffusion of the two surfaces in contact, temperature and time of contact, the nature and thickness of oxide films or contaminants present at the interface, and the presence of a lubricant film.

It is apparent that, when sliding the two surfaces along each other under an external tangential force, these junctions must be broken by shearing them. This shearing force is the friction force. Since the coefficient of friction μ is defined as the ratio of the friction force F to the normal force N, we note that

$$\mu = \frac{F}{N} = \frac{\tau A_r}{\sigma A_r} = \frac{\tau}{\sigma} \tag{5.1}$$

where τ is the shear stress, σ is the normal stress at the junction, and A_r is the real area of contact.

This equation is significant in that it illustrates the importance of the strength of the interface on friction. It can be seen that μ is reduced by lowering the interfacial shear strength. The denominator in this expression is basically the yield strength of the asperity, and its magnitude depends on the particular material. It is not affected to any significant extent by the presence of contaminants, oxides, or lubricants at the interface, although it is of course affected by temperature.

With clean nascent surfaces (such as those produced by cutting, or in forming operations in which surface extensions are large) and in the absence of any contaminants or lubricant film, the strength of the junctions is high because of cold pressure welding. Consequently, the shear strength of the junction is high, and hence friction is high. As contaminants or lubricants are introduced, or as oxide layers develop (which may take only a few seconds in some cases), the strength of the junction is lowered because, under these conditions, a strong bond cannot be formed. Thus, friction is lower.

Another way of characterizing friction is to define a friction factor (or shear factor) m as

$$m = \frac{\tau_i}{k} \tag{5.2}$$

where τ_i is the shear strength of the interface and k is the shear yield stress of the softer material in the sliding pair. Thus, when m = 0, friction is zero, and when m = 1 complete sticking takes place. Sticking does not necessarily mean that complete adhesion occurs at the interface. It indicates that the shear strength at the interface has reached the shear yield stress k of the material; in this case the softer material, since it is the softer material that shears more easily.

In addition to the mechanism of shearing at the junctions, friction may also be due to ploughing (or plowing). A hard asperity scratches the softer surface (with or without a chip or sliver being produced on the softer surface) during sliding (Fig. 5.2). Since this involves plastic deformation, the energy required is supplied by the ploughing force. It has been shown that the ploughing force can contribute significantly to the friction force.

Friction force raises the temperature at the interface. The temperature increases with sliding speed, coefficient of friction, and decreasing thermal conductivity and specific heat of the materials. The higher the thermal conductivity, the greater is the heat conduction into the bulk of the workpiece. Also, the higher the specific heat, the lower is the temperature rise. Temperatuer rise can be sufficiently high to melt the interface or to cause phase transformations, residual stresses, and surface damage (metallurgical burn).

It can be seen from the foregoing description that the function of a lubricant is primarily to reduce the shear strength at an interface, that is, lower the value of the numerator in Eqs. (51.) and (5.2). This is carried out through various mechanisms, described in Sec. 5.3. (A lubricant has additional functions, such as cooling the interface, protecting newly formed surfaces against environmental attack, and washing away chips in machining operations.)

5.2 WEAR

Wear is defined as the loss or removal of material from a surface. Wear can take place under different conditions. These may be dry or lubricated wear, sliding or rolling contact wear, and wear by fracture or by plastic deformation.

There are four basic types of wear: adhesive, abrasive, fatigue, and corrosive wear. Generally, the first three types are of interest in metalworking operations, although the last can also occur as a result of tool, die, and workpiece interactions in the presence of various liquids and gases. Particularly in this case, appropriate choices in lubricant chemistry must be made, depending upon the tool and die compositions, Otherwise, excessive corrosive tool wear takes place.

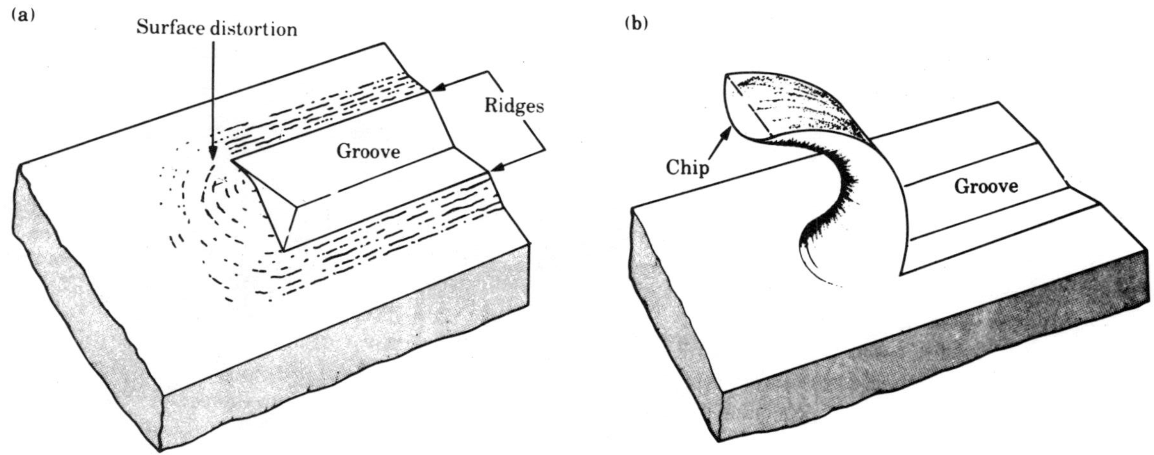

Figure 5.2 (a) Ploughing on a surface, and (b) chip formation with an abrasive. (After L. E. Samuels. From S. Kalpakjian, 1984. *Manufacturing Processes for Engineering Materials,* Addison-Wesley, Reading, Massachusetts, p. 199.)

5.2.1 Adhesive Wear

This type of wear is a result of the junctions sheared during sliding. If the junctions have strong bonds (such as with clean interfaces, under high loads, and with sufficient time of contact between the two bodies), fracture of the junction takes place either above or below the interface of the asperities. Generally, it is through the softer metal that a crack forms and propagates. Under repeated cycling, the transferred particle becomes a loose wear particle. In severe cases of adhesive wear, the process is called galling, scuffing, or seizure.

Based on the probability that a junction would lead to a wear particle formation, an expression for adhesive wear has been given in the form of

$$V = k\ \frac{LW}{3p} \tag{5.3}$$

where V = volume of material removed, k = wear coefficient, L = length of travel, W = normal load, and p = indentation hardness of the softer body. Various tables are available in the literature on the k values for several materials, obtained experimentally under vaious conditions.

For adhesive wear to occur, there must be an affinity (reactivity) for adhesion and welding between the two sliding surfaces. The most severe case of wear occurs between two clean surfaces, under high normal load and in vacuum. The basic role of an effective lubricant is to reduce the tendency for welding of the asperities, either by separating the surfaces with a layer of lubricant or by reducing the shear strength of the interface by forming low-shear-strength compounds through chemical reactions.

Surface films are of great importance in adhesive wear. Other than the lubricant layer, the surfaces are almost always covered with oxide layers, contaminants, and adsorbed gases or fluids. These films significantly reduce the shear strength of the interface. Thus, the wear observed in practice is generally lower than it would have been otherwise. Oxide films have a significant role in friction and wear. Their effect depends on the relative rate at which oxide layers are destroyed during sliding and the rate at which they form. If the rate of destruction is high, the surfaces are not as well protected, and wear is high.

5.2.2 Abrasive Wear

In this wear process, material is removed from the surface by scratching and producing slivers or microchips. Thus, the softer the material, the higher is the abrasive wear rate. Also, the higher the load, the higher is the wear rate.

Abrasive wear may be of the two-body and three-body type. In the latter, the third body is composed of wear particles or any other hard contaminants (such as those built up in a lubricant) that are trapped between the two sliding bodies. This mechanism may also be referred to as erosive wear. This type of wear is important in metalworking processes and in the maintenance of equipment. To reduce the buildup of oxides, metal chips, or other metallic particles, periodic inspection, filtering, or changing the lubricants is therefore good practice.

5.2.3 Fatigue Wear

This type of wear is generally called surface fatigue or surface fracture wear. It is due to cyclic loading of an interface between the tool or die and the workpiece. Cracks develop on the surface over a period of time by a fatigue mechanism, due either to mechanical forces or to thermal stresses (thermal fatigue). In either case, material is removed from a surface (usually the tool or die) by spalling or pitting, whereby cracks coalesce by joining each other below the surface.

The role of lubrication in fatigue wear is complex. Lubricants reduce friction and hence reduce the level of stresses that could cause fatigue failure. On the other hand, if a crack develops due to some mechanism or cause, the fluid penetrates the crack by surface tension. During subsequent loading cycles the fluid is trapped and, because it is incompressible, develops high hydrostatic pressure in the crack opening. This, in turn, propagates the crack farther into the body of the tool or die. Pitting, for example, does not occur in unlubricated interfaces unless chemical attack takes place.

5.3 LUBRICATING MECHANISMS

It is apparent from the foregoing discussions that friction and wear can be reduced or eliminated by keeping the sliding surfaces apart from each other. Whereas in machine elements, such as lubricated journal bearings and air bearings, this requirement may be fulfilled easily, the loads and speeds involved in metalworking and the geometry of the tool, die, and workpiece interfaces are generally such that they do not readily permit the existence of a lubricant film. In this section the major lubricating mechanisms of interest to metalworking processes are described.

5.3.1 Thick-Film (Hydrodynamic) Lubrication

In this regime of lubrication (also called full-fluid film), the two surfaces are completely separated from each other by a continuous fluid film. The thickness of this film is about 10 times the magnitude of the surface roughness of the mating surfaces. The fluid film can be developed either hydrostatically (by entrapping the lubricant) or, more generally, by the wedge effect of the sliding surfaces in the presence of a viscous fluid at the interface. Thus, in this type of lubrication, the bulk properties of the lubricant (especially viscosity) are important; chemical effects of the lubricant on metal surfaces are not significant.

In thick-film lubrication the loads are usually light and the speeds are high. The coefficient of friction is very low, generally between 0.001 and 0.02. There is no wear, except due to any foreign matter (third body) that may have entered the lubricating system. This type of lubrication does not generally occur in metalworking processes, except in isolated regions at die-workpiece interfaces with high-viscosity lubricants and at high operating speeds. Some examples are upsetting at high speeds with a viscous lubricant and hot extrusion with glass as the lubricant. Forming at very high pressures, such as in hydrostatic extrusion, may also involve this type of lubrication.

5.3.2 Mixed Lubrication

The film thickness in thick-film lubrication can be reduced by decreasing the viscosity (such as due to temperature rise), decreasing the sliding speed, or increasing the load. The surfaces become close to each other and the normal load between the tool or die and the workpiece is supported partly by the fluid film in hydrodynamic pockets in the surface roughness of the interfaces and partly by metal-to-metal contact of the surfaces. This is generally referred to as mixed lubrication and also as the thin-film or quasi-hydrodynamic regime.

The film thickness is less than three times the surface roughness; the coefficient of friction may be as high as about 0.4 (thus forces and power consumption may increase substantially), and wear can be significant. There is an optimum roughness for effective lubricant entrapment, with a recommended roughness of usually 15 μin. (0.40 μm). The hydrodynamic pockets also serve as reservoirs for supplying lubricant to those regions at the interface that are starved for lubricants.

5.3.3 Boundary Lubrication

In this type of lubrication a thin layer of lubricant film physically adheres to the surfaces by molecular forces, such as van der Waals forces, or by chemical forces (chemisorption). Typical boundary lubricants are oils, fatty oils, fatty acids, and soaps. Boundary films can form rapidly on clean surfaces, although reactivity on some metals, such as titanium and stainless steel, is very low; lubri-

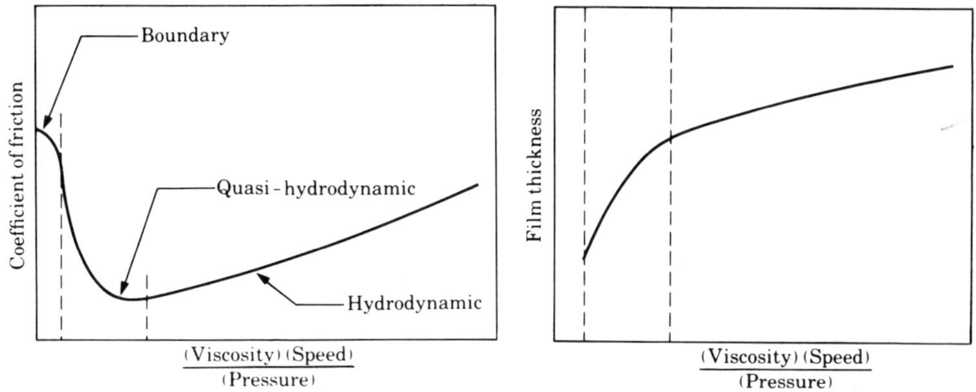

Figure 5.3 Stribeck curve showing various lubrication regimes as a function of viscosity of the lubricant, sliding speed, and interface pressure. (S. Kalpakjian, 1984. *Manufacturing Processes for Engineering Materials,* Addison-Wesley, Reading, Massachusetts, p. 218.)

cation may then be enhanced by the formation of boundary films on tool and die surfaces instead of the surface of the workpiece.

An important distinction is that, unlike in full-fluid film lubrication, where the bulk properties of the lubricant, such as viscosity, are important, in boundary lubrication the chemical aspects of the lubricant and its reactivity with metal surfaces are important; viscosity has a secondary role. In the boundary lubrication regime, the coefficient of friction usually ranges between 0.1 and 0.4, depending on the strength and thickness of the boundary film. Boundary lubrication is often observed and practiced in metalworking operations.

Wear rate in this regime depends on the rate at which films are destroyed by rubbing off, or by desorption due to excessive temperatures generated during the metalworking process. If the protective boundary layer is destroyed, friction and wear will be high. The adherence and strength of this film is, therefore, a very important factor in the effectiveness of boundary lubrication. The role of pressure, speed, and viscosity on film thickness should also be recognized, as indicated by the well-known Stribeck curve in Fig. 5.3.

5.3.4 Extreme-Pressure Lubrication

In this regime the surface of the metal is altered chemically by irreversible chemical reactions. These reactions, involving sulfur, chlorine, and phosphorus in the metalworking fluid, form salts of the mating metal surfaces. These surfaces prevent or reduce welding of the asperities at the interface even under high die-workpiece contact pressure, hence the term "extreme pressure" (EP). Furthermore, because of their low shear strength, these surface films also reduce friction.

As temperature increases, however, these films may break down, the temperature for breakdown depending on the particular EP additive (used either singly or in combination, such as both sulfur and chlorine) and the composition of the metal surfaces. When the film breaks down, metal-to-metal contact takes place, with a subsequent increase in friction and wear. However, the protective films of sulfides and chlorides form again with relative ease, especially on clean new surfaces. Air, oxygen, humidity, and water also have important effects on extreme-pressure lubrication.

5.3.5 Elastohydrodynamic and Plastohydrodynamic Lubrication

It will be noted that no mention has been made regarding the deflections and distortions the tools and dies may undergo due to the stresses encountered in metalworking operations. It has been shown that these deflections, due to the finite modulus of elasticity of materials, can be sufficiently extensive to alter the geometry of the die-workpiece interface, thus affecting the stresses, contact area and geometry, and pressure distribution. Hence the term "elastohydrodynamic" (EHD). Another factor is the increase in viscosity (and even solidification) of lubricants with pressure. This, in turn, helps develop hydrodynamic films, resulting in an increase in film thickness.

An extension of EHD is plastohydrodynamic lubrication (PHD). In this regime, encountered in such processes as drawing and strip rolling, the lubricant is entrained or entrapped at the converging gaps in the die-workpiece interfaces. Thus, a full fluid film is developed with a large drop in friction and wear.

These phenomena are particularly important in processes in drawing and in concentrated contacts, such as cold rolling of thin strip, because of the influence of small changes in the relative interfacial dimensions on forces and deformation geometry.

5.3.6 Solid-Film Lubrication

In this type of lubrication, layer-lattice compounds, principally graphite and molybdenum disulfide (MoS_2), are deposited on the surfaces by "running in" mixed with a fluid carrier, such as water or oil, or with a polymeric binder. Solid lubricants, because of their layer structure, have low shear strength and hence reduce friction at the interface. However, their shear strength depends on the environment. Thus, graphite is useful in an environment of water vapor or oxygen. In vacuum its friction is high and it can, in fact, become abrasive. Conversely, MoS_2 is more effective in vacuum than in air.

New developments in solid lubrication include various polymers. In the well-known case of PTFE (Teflon), the low friction observed has been attributed to the transfer of a polymer film to the metal surface. The natural oxides on metal surfaces may also act as lubricants, depending on the type of material and the nature of the oxide layer. (See also Chap. 6.)

For solid lubricants to be effective, they must cover the surfaces with a continuous film; otherwise, the high spots at the interface may be starved for lubricants and metal-to-metal contact may take place. The strength of the solid film depends on the method of application and the nature of the reaction at the lubricant-metal interface. Solid lubricants are particularly useful at elevated temperatures.

5.4 GEOMETRIC EFFECTS

From the analysis of bearings and sliding systems for machine elements, it can be recognized that the geometry of the tool, die, and workpiece interfaces is an important aspect of lubrication in metalworking. Particularly significant is the region where the workpiece enters the die, such as in rod drawing or rolling. The entrainment or entrapment of lubricants, to carry them into the interfaces, can be improved by using smaller wedge angles at the entrance; otherwise, the lubricant tends to be scraped off.

On the other hand, it may not always be desirable to have a thick layer of lubricant at the die-workpiece interfaces. If the surfaces are separated completely, the surface of the deformed workpiece maintains a dull appearance, not having come in contact with the die.

Some of the effects of surface roughness in lubrication are discussed in Chap. 4. Although die surfaces should generally be smoother than the workpiece surface (especially in the mixed-film regime), it has been shown that a certain optimal roughness of the die surface is desirable to entrap solid lubricants. Although these lubricants are also entrapped on the workpiece surfaces, thus making it an effective means of lubrication, this situation leads to the depletion of the lubricant as the part is made and removed from the machine.

5.5 ROLE OF PHYSICAL PROPERTIES OF THE LUBRICANT

From the foregoing discussions we have seen that, in general the bulk properties of the lubricant (viscosity) and its chemical properties in relation to the workpiece and die materials are major factors in effective lubrication in metalworking. In addition to the properties previously stated, other physical properties also play an important role in lubrication. These are surface tension and wetting. Wetting is a phenomenon related to surface tension, which is a manifestation of surface energy.

Wetting characteristics of a lubricant, that is, how well the lubricant spreads itself over the surface of the workpiece as a continuous film, is an important aspect of lubrication. There may be situations, however, in which it is desirable for the lubricant to remain in a certain area of the tool-workpiece interface. Watch manufacturers, for example, have recognized the need for nonmigrating (nonwetting) lubricants for pivot points in a watch. In some metalworking operations, such as forging and deep drawing, there are regions where the lack of a lubricant improves the operation. In forging, this situation aids in die filling, and in deep drawing it aids in drawability by having the cup travel with the punch.

The shape of a drop of fluid (such as metalworking lubricant) on a solid metal surface depends on the interfacial tensions between the metal, fluid, and air. The angle that the periphery of the droplet makes with the surface is called the contact angle. The smaller this angle, the better are the wetting characteristics of the fluid.

Wetting in metalworking fluids is improved by the addition of wetting agents, such as alcohols and glycols, or by increasing the temperature. It is also found that wetting is improved by increasing the surface roughness.

It can be seen from the foregoing discussion that lubrication in metalworking involves different mechanisms, depending upon the chemistry of the tool-lubricant-workpiece interface, the method of lubricant application, and the geometry of the process, as well as the mechanics of the operation. Furthermore, the lubrication mode often varies during the metalworking cycle, depending upon the changes in the speed of the process as well as the amount of deformation and attendant pressures and stresses involved.

BIBLIOGRAPHY

Bartz, W. J., and J. Wolff, eds., 1982. *Lubrication in Metal Working,* 2 vols., Technische Akademie Esslingen, Germany.

Bastian, E. L. H., 1951. *Metalworking Lubricants,* McGraw-Hill, New York.

Billet, M., 1979. *Industrial Lubrication,* Pergamon, New York.

Bowden, F. P., and D. Tabor, 1974. *Friction: An Introduction to Tribology,* Heinemann, London.

Braithwaite, E. R., ed., 1967. *Lubrication and Lubricants,* Elsevier, London.

Cameron, A., 1980. *The Principles of Lubrication,* 2d ed., Longman, London.

Clauss, F. J., 1972. *Solid Lubricants and Self-Lubricating Solids,* Academic Press, New York.

Cutting and Grinding Fluids, 1967. Society of Manufacturing Engineers, Dearborn, Michigan.

Czichos, H., 1978. *Tribology: A Systems Approach,* Elsevier, Amsterdam.

Friction and Lubrication in Metal Processing, 1966. American Society of Mechanical Engineers, New York.

Halling, J., 1975. *Introduction to Tribology,* Macmillan, London.

Hondros, E. D., 1971. *Tribology,* Mills and Boon, London.

Kalpakjian, S., and S. Jain, eds., 1980. *Metalworking Lubrication,* American Society of Mechanical Engineers, New York.

Kragelskii, I. V., 1965. *Frication and Wear,* Butterworth, Washington, D.C.

Lansdown, A. R., 1982. *Lubrication – A Practical Guide to Lubricant Selection,* Pergamon, New York.

Lubrication Challenges in Metalworking and Processing, First International Conference, 1978; Second International Conference, 1979. IIT Research Institute, Chicago.

Moore, D. F., 1975. *Principles and Applications of Tribology,* Pergamon, New York.

Neale, M. J., ed., 1973. *The Tribology Handbook,* Butterworth, London.

O'Connor, J. J., et al., eds., 1968. *Standard Handbook of Lubrication Engineering,* McGraw-Hill, New York.

Peterson, M. B., and W. O. Winer, eds., 1980. *Wear Control Handbook,* American Society of Mechanical Engineers, New York.

Pugh, B., 1973. *Friction and Wear: A Tribology Text for Students,* Newnes-Butterworth, London.

Rabinowicz, E., 1965. *Friction and Wear of Materials,* Wiley, New York.

Rigney, D. A., ed., 1981. *Friction and Wear of Materials,* American Society for Metals, Metal Park, Ohio.

Sarkar, A. D., 1976. *Wear of Metals,* Pergamon, New York.

Schey, J. A., ed., 1970. *Metal Deformation Processes: Friction and Lubrication,* Marcel Dekker, New York.

Schey, J. A., 1983. *Tribology in Metalworking: Friction, Lubrication and Wear,* American Society for Metals, Metals Park, Ohio.

Suh, N. P., and N. Saka, eds., 1980. *Fundamentals of Tribology,* MIT Press, Cambridge, Massachusetts.

Tribology in Metal Working – New Developments, (no date). Mechanical Engineering Publications, London.

6
Chemistry of Metalworking Lubricants

Lubricants in metalworking processes are used primarily to interpose a film between the tool or die and the metal being formed or machined. This film minimizes adhesion or welding of one surface to the other. There are, however, other important functions, as described in Chap. 5 and outlined in Table 6.1.

Tool and die life and part integrity are significantly influenced by the adequacy of lubrication. The lubricant film varies in thickness and structure, depending upon the process conditions (temperature, pressure, and geometry) and the chemical and mechanical properties of the lubricating film, as well as the surface characteristics of both the tool and the workpiece. All these factors combine to influence film performance as a lubricant (Fig. 6.1). Subsequently, the chemistry of the film influences the likelihood of corrosion of the tool and metalworking equipment, as well as of the parts; it also affects welding, painting, efficiency and cost of cleaning, recycling of the used fluid, and disposal costs. The chemistry of the lubricant also influences the likelihood of biological degradation, as well as the possibility of the occurrence of operator dermatitis. Thus, the chemistry of lubricants has broad technological, health, and cost implications for the total metalworking operation.

In this chapter many of the more important chemical aspects of metalworking lubrication are described. In an attempt to impose some order on the ensuing presentation, several somewhat arbitrary decisions have been made with respect to the subject matter. This chapter is divided into five main sections: (1) water-based systems, (2) oil-based systems, (3) synthetic systems, (4) Suspensions, pastes, gels, salts, and solid films, and (5) additive

Table 6.1 General Characteristics of Metalworking Fluids[a]

Function	Compounded oil	Oil-in-water emulsion	Semi-synthetic solution	Synthetic solution
1. Provide lubrication at high pressure (boundary lubrication)	1	2	3	4
2. Reduce heat from plastic deformation (heat transfer)	5	2	2	1
3. Provide cushion between workpiece and die to reduce adhesion and pickup (film thickness)	1	2	3	4
4. Reduce friction between die and workpiece and, thus, heat	1	2	2	1
5. Reduce wear and galling between tool and workpiece (chemical surface activity)	4	3	2	1

6. Flushing action to prevent buildup of scale, dirt (fluid flow)	5	4	3	2
7. Protect surface characteristics; nonstaining	5	2	2	1
8. Minimize processing costs, welding, cleaning, painting	5	2	2	1
9. Minimize environmental effect, air contamination, water recovery	5	2	2	1

[a]*Note:* 1 = most effective, 5 = least effective.

Source: After E. S. Nachtman, 1980. In *Metalworking Lubrication,* S. Kalpakjian and S. C. Jain, eds., American Society of Mechanical Engineers, New York, p. 53.

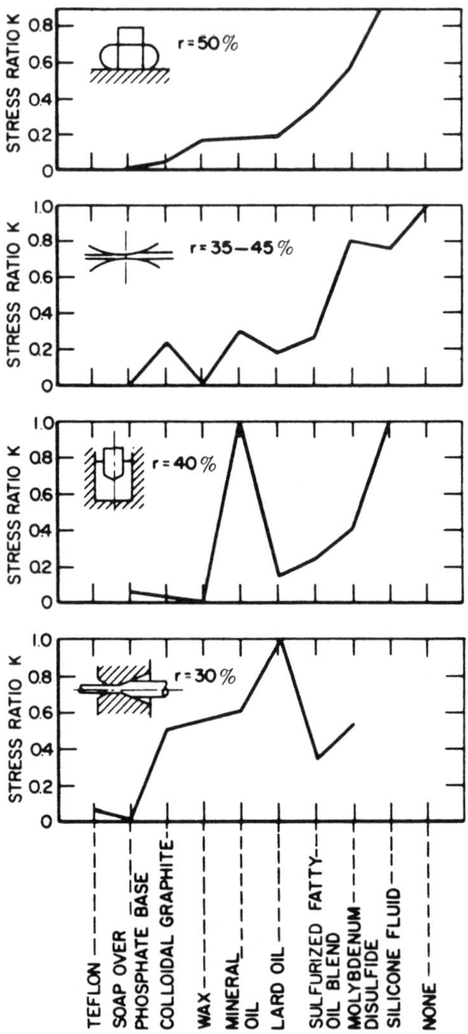

Figure 6.1 Relative performance of various lubricants in (a) wire drawing, (b) piercing, (c) flat rolling, and (d) upsetting. (Development of Method of Evaluating Lubricants for Cold Working of Metals, Final Report, WAL-TR-364/35, July 1959, Massachusetts Institute of Technology, Boston). (From J. A. Schey, ed., 1970. *Metal Deformation Processes: Friction and Lubrication,* Marcel Dekker, New York, p. 293.)

chemistry. There is an unavoidable overlap between these sections, but in an attempt to clarify as much as possible of lubricant chemistry, this organization has been adopted.

6.1 WATER-BASED SYSTEMS

Metalworking lubricants consisting of water-miscible (water-soluble) constituents and water-based emulsions have become, with increasing frequency, the lubricant of choice. This growth in use has accelerated in recent years due to the increasing cost of oil-based lubricants and the concurrent development of new chemical systems. The developing sophistication of the chemical components of water-based lubricants has resulted in improved metal wetting and corrosion protection properties, thus making them attractive alternatives to oil-based lubricants for many operations previously only satisfactorily lubricated with oil-based lubricants.

Furthermore, water-based lubricants may have a variety of physical properties, depending upon whether they are emulsions, solutions, suspensions, gels, or pastes. Water may be common to all these states, and hence, it is appropriate to start this discussion of water-based systems by first describing the chemistry of water.

6.1.1 Chemistry of Water

The formula for water (H_2O) gives us information about its composition and molecular weight but does not enlighten us with respect to its structure. Indeed, the structure of water is surprisingly complex. Water consists of some combination of reactive monomers associated with nonreactive clusters of water monomers.

The reactive water molecule is asymmetric; it is positively charged on the hydrogen end and negatively charged at the oxygen end. This asymmetry in bonding energy leads to agglomeration, with the hydrogen of one molecule attracted to the oxygen of another. This attractive force (hydrogen bonding) leads to the cluster formation postulated as occurring in pure water.

One of the practical consequences of this bonding is manifest in the reluctance of water molecules to evaporate from the surface of a body of water as rapidly as would be the case if this bonding phenomenon did not occur. The hydrogen bond is strong and difficult to rupture. This high energy, as compared with that of other fluids, directly results in a high energy of vapor formation or heat capacity. This property of water means excellent heat-transfer characteristics. These properties of high heat transfer and heat capacity are used to advantage in the lubrication of metalworking operations, since heat and temperature control may be critical in achieving the desired operational results.

Water is often called the universal solvent since so many ionic, as well as nonionic, substances are solubilized by water. Water molecules in contact with a crystal orient themselves relative to the crystal to reduce or eliminate the bonding forces holding together the atomic structure of the crystal. The resulting ions may be hydrated by the water molecules contacting the crystal, thus changing the bonding energy and preventing recombination or recrystallization of the original crystalline structure.

This solvency and hydration effect is shown quantitatively by the relatively high dielectric constant of water. Water ionizes very slightly, producing only 10^{-7} mol hydrogen and 10^{-7} mol of hydroxyl ions per liter; thus, it is an insulator. When salts or other ionizing materials are dissolved in water, conduction of an electrical current may occur. Thus, the conductivity of naturally occurring waters provides a measure of the dissolved mineral content. This property is taken advantage of in tests designed for control of water-based metalworking fluids.

The viscosity of water affects its treatment and use. Viscosity is a measure of internal friction, the friction of one layer of molecules moving relative to another. As the temperature of water is increased, internal friction decreases. Dissolved salts and gases can diffuse more rapidly in water as the temperature rises, and the physical processes of sedimentation and degasification occur at a high rate.

All naturally occurring water is impure. Solvent action of water varies from one mineral to another and from one compound to another. The precipitation of most inorganic solids from water is so affected by the formation of ion pairs and complexes that the solubility of a specific precipitate, because of the great variety of ions present in most water systems, is difficult to predict. Furthermore, some cations, such as lithium or aluminum, have such a strong ionic attraction for the water molecule that very large clusters of hydrated aluminum or silicon complexes may form and, in turn, influence solutions or emulsions formulated for metalworking.

Fundamental to an understanding of the effects of water on lubricants is the hardness of water. Traditionally, hardness is defined as the soap-consuming capacity of water. Hardness, then, is directly related to the concentration in water of calcium, magnesium, and other heavy metals, such as iron and manganese, independent of the nature of anions present. Lesser constituents are often ammonia, borate, fluoride, iron, nitrate, potassium, strontium, aluminum, arsenic, barium, phosphate, copper, and zinc. Furthermore, nonsoluble constituents may be suspended as solids, and microbial organisms, such as algae, bacteria, fungi, and viruses, are all, to some degree, always present.

The hydrogen ion concentration can be measured with a pH meter. It also can be titrated when the concentration becomes high enough to be detectable by chemical analysis. Since the pH is a logarithmic function, the hydrogen ion concentration increases by factors of 10 for each unit of pH reduction. Measurement of pH can be very valuable in determining the hydrogen and hydroxyl ion concentration, often a prerequisite for control of water-based metalworking fluids.

Systems for controlling the cationic concentration (Ca, Al, Mg, and Fe) in water may be critical in terms of water-based lubricant performance. Deionization or precipitation and coagulation of these ions are the preferred methods for controlling their concentration. Water-softening systems are much less effective, substituting sodium for other cations, but not affecting the overall cationic concentration.

The source of water used for the makeup of the concentrate for the diluted metalworking lubricant is also important. The chemistry, and particularly the hardness, may vary widely, depending upon whether the water is obtained from deep wells, city treatment plants, lakes and streams, boiler water, or rainwater. Furthermore, seasonal variations from the same source can also influence the chemistry and composition of the water. An awareness of these source and seasonal variations can be very helpful in anticipating operating problems. Thus, the formulator must take into account the chemistry, as well as other properties of the water, since the characteristics we have been discussing influence the chemistry of additives and their performance.

Hard water, for example, can result in destabilization of an emulsion, since the cations, such as calcium, magnesium, and aluminum, will inactivate some or all of the surfactants used in the formulation. Rust preventative additives may similarly be reduced in their effectiveness. Biocides may be complexed and rendered impotent or, more frequently, bacteria and fungi are nourished and grow, feeding on the iron, manganese, and other cationic complexes in the water. Uncontrolled bacterial growth and unexpected corrosion, as well as poor lubrication, may occur because of an insufficient awareness of the importance of the chemistry of water.

Let us now turn our attention to consideration of the chemistry of emulsions and solutions, the most widely used water-based systems.

6.1.2 Emulsions

An emulsion may be defined as a system consisting of at least one immiscible liquid dispersed in another in the form of droplets. Such systems have limited

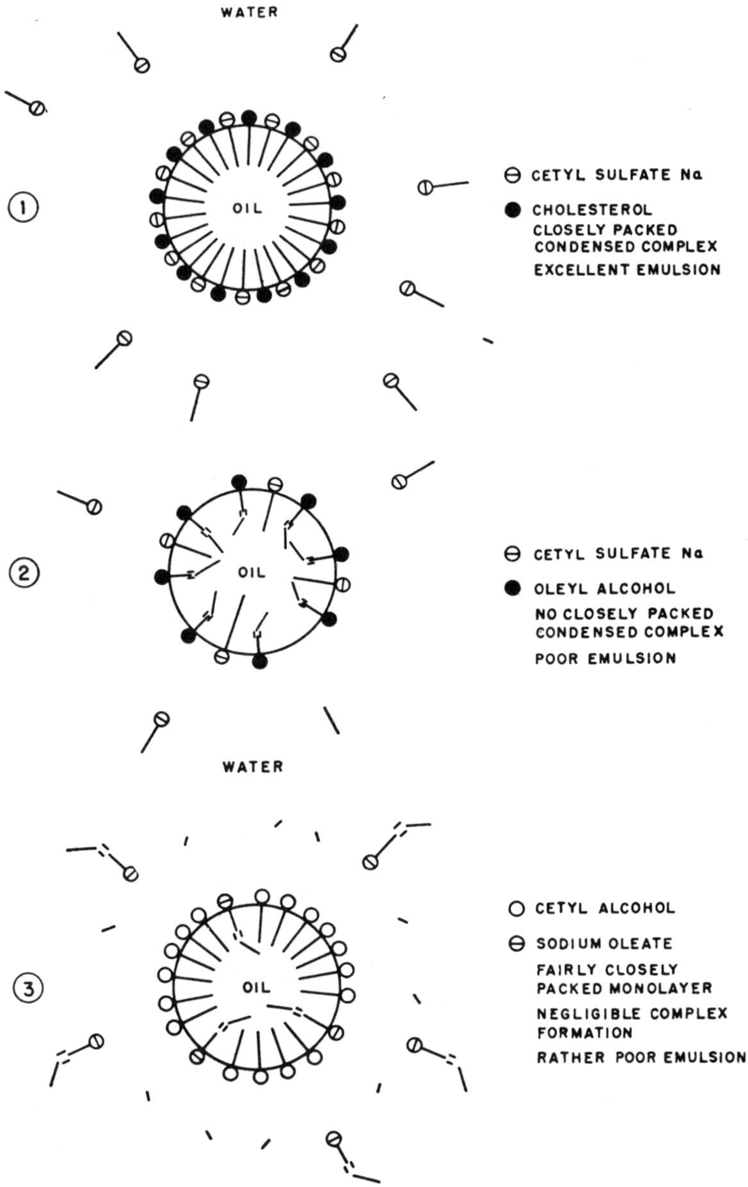

Figure 6.2 Complex formation at the oil-water interface in emulsions. (J. H. Schulman and E. G. Cockbain, 1940. *Transactions of the Faraday Society,* 36, p. 651.)

stability. Therefore, such additives as surface-active agents and finely divided solids, as well as processing methods, such as homogenization (a mixing technique), may be used to improve stability. The phase present in the form of finely divided droplets is generally called the internal phase, or dispersed phase. The matrix in which these droplets are suspended is called the external phase, or the continuous phase.

The size of the dispersed globules or globlets suspended in the continuous phase governs the appearance of the emulsion. If the diameter of the dispersed droplet is 1 μm or larger, the emulsion is milky white in color. If the diameter is less than 1 μm but greater than 0.1 μm, the emulsion appears to be blue-white in color. If the droplet diameter varies between 0.1 and 0.05 μm, the emulsion looks gray and is semitransparent. If the droplets are less than 0.05 μm in size, the emulsion is clear and transparent.

For the droplets of the dispersed phase to be sufficiently stable for the two-phase system to be called an emulsion, a third component must be added, that is, an emulsifier or surfactant. Fine solid particles and the mixing practice may also influence emulsion stability. The most widely applicable technology used for the formation of metalworking emulsions utilizes emulsifiers (Fig. 6.2).

Two distinct emulsion states are used for metalworking lubricants. The first, and by far the most commonly used, consists of an emulsion of oil in droplet form as the internal phase, dispersed in water as the continuous, or external, phase. Oil, in the course of this discussion, may refer to mineral or vegetable oil or a synthetic organic molecule in fluid form. The most commonly used oil is a refined mineral oil. Less commonly used is the invert emulsion, which consists of water droplets dispersed in the continuous phase of oil (Fig. 6.3).

Multiple emulsions have been used in a few metalworking applications. These consist of systems in which oil-in-water as well as water-in-oil emulsion states occur in the same system at the same time. What is observed is an oil droplet that contains water enclosed by a continuous phase consisting of water; this alternation of state may then be multiplied.

The difference between an oil-in-water and water-in-oil emulsion is readily determined. If water is added to the dilute emulsion, the oil in water is easily diluted. The water-in-oil emulsion will, only with difficulty if at all, accept the water. On the other hand, if oil is added to the same two systems, the oil-in-water emulsion accepts the oil with difficulty, if at all, and the water-in-oil emulsion is readily diluted. The conductivity of oil-in-water emulsions is similar to that of water. There is little, if any, water conductivity in a water-in-oil emulsion. Finally, an oil-soluble dye will color a water-in-oil emulsion

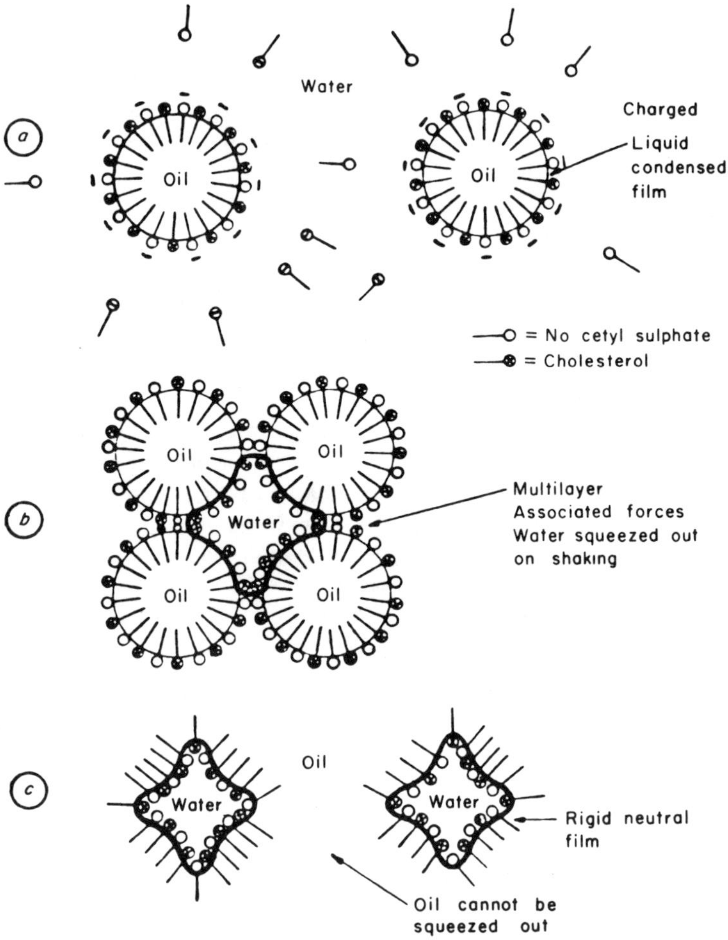

Figure 6.3 Inversion of an oil-in-water emulsion. (J. H. Schulman and E. G. Cockbain, 1940. *Transactions of the Faraday Society,* 36, p. 61.)

readily and give almost no color in an oil-in-water emulsion. The reverse phenomenon is observed if a water-soluble dye is added to the emulsion system.

In the case of lubricants used in the form of emulsions, the importance of surface activity is overriding. If, for example, one considers a given volume of oil compared with the same volume in the form of a 1-μm globule suspended in water, the increase in total interfacial area is on the order of 100,000 times. The effect of the increase in surface area amplifies the importance of interfaces and the reaction that may occur there.

Interfacial Phenomena

Additional phenomena that occur in both emulsions, as well as solutions, deserve some discussion since they relate to metal wetting as well as film strength, both of which are important in metalworking lubrication. These phenomena are adhesion and cohesion, on the one hand, and surface tension on the other, as well as the relationship between bulk and surface properties in terms of micelle formation.

To understand the relationship between surface and interfacial tension, the concepts of adhesion and cohesion must be reviewed briefly. If a liquid of a given composition can be considered in a solid form in the shape of a bar subjected to opposing forces pulling at the extreme ends, as in a tensile test until fracture occurs, the force required for fracture is a measure of the work of cohesion (film strength). Now, on the other hand, consider a liquid in contact with and floating on the surface of steel. Through some appropriate device a force is then applied to the liquid to the point where the liquid parts company with the steel surface. This parting force is a measure of the work of adhesion. The relationship between these forces with respect to the interfaces between liquids, solids and liquids, and gases and liquids, as well as the forces of cohesion within the liquid, solid, or gas, determine the performance of the lubricant film.

To illustrate, fatty acids are common emulsifying agents used to form emulsions for metalworking. They have a profound effect on the surface tension of water. This effect is strongly dependent upon their molecular weight. Higher molecular weights have a greater effect on lowering the surface tension than an equal concentration of a lower molecular weight fatty acid. This major reduction in surface tension results from the fact that the fatty acids acting as surfactants concentrate at the surface.

Molecules of this type have been given the designation amphipathic, that is, a molecule that has both water-loving (water-soluble) and water-hating (water-insoluble) characteristics. In the case of a fatty acid, that portion of the structure consisting of the carbon-hydrogen chain is hydrophobic (water hating) and enters the oil phase of the emulsion. The acid component of the molecule, on the other hand, is hydrophilic (water loving) and enters the water phase, all of this occurring at the interface between oil and water.

It was pointed out previously that surfactants can greatly decrease the amount of energy required to create new surface area. They do this by reducing the interfacial tension, or the surface free energy, of the interface between the two liquids to be emulsified. The surface free energy for an interface is equal to the sum of the surface free energies of the two liquids involved minus the interaction energy (or attractive forces) between surface molecules of the adjacent liquids. The greater the similarity between the two liquids in structure, the greater is this interaction, and therefore, a lower interfacial tension will result.

Consider, then, what happens when a suitable surfactant is added to a pair of immiscible liquids, such as oil and water. The surfactant will be absorbed at the interface between the two liquids and it will orient itself; the hydrophilic group toward the water and the lipophilic (oil miscible) toward the oil. Where the surfactant molecules replace water and/or oil molecules of the original interface, the interaction across the interface is now between the hydrophilic group of the surfactant and water molecules on one side of the interface and between the lipophilic group of the surfactant and oil on the other side.

Since these interactions are now much stronger than the original interaction between the highly dissimilar oil and water molecules, the tension across the interface is significantly reduced by the presence of the surfactant. Thus, less energy will be needed to form an emulsion of the oil and water phases.

Surfactants

A large number of surfactants is available. Suitable choices are based upon economics as well as on the chemical and physical characteristics desired. Depending upon the nature of the hydrophilic group, surfactants are classified as anionic (negative charge), cationic (positive charge), zwitterionic (both positive and negative charges are present), and nonionic (no apparent ionic charge is present).

The hydrophobic portion of the surfactant also varies in chemistry and properties, but not as significantly. Straight or branched-chain hydrocarbons or combinations of both are commonly used. This two-headed entity, one head hydrophobic and the other hydrophilic, is selected to provide the coupling interaction between the oil and water phases, to create the desired chemical and physical properties for the lubricant.

The versatility of the surface activity occurring at the oil-water interface is associated with still another phenomenon, micelle formation, which reflects the properties and compositional changes brought about in formulating lubricant emulsions (Fig. 6.4). Solubilization in solutions is also influenced by micelle formation and is now discussed in that regard in connection with water solutions as metalworking lubricants.

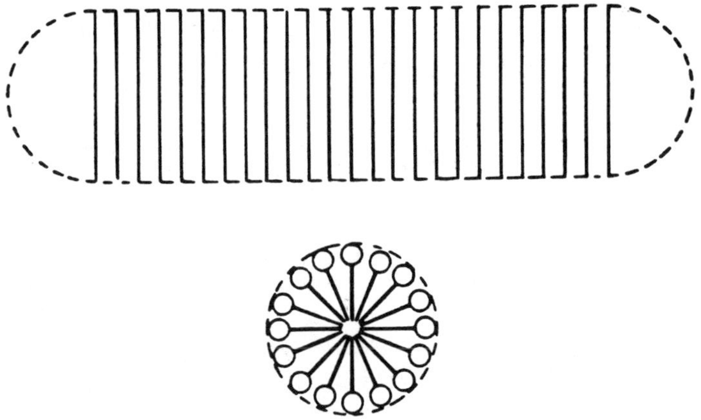

Figure 6.4 Schematic illustration of the Debye rodlike micelle.

The theory of micelle formation in emulsions and solutions is based on the spontaneous formation from surface-active solutes of clusters of colloidal dimensions from ions; that is, the colloidal-sized clusters are in true reversible equilibrium with the ions. These stable colloidal particles have a self-organizing structure in which interior regions, containing the hydrophobic groups of the surfactant molecules of a radius comparable to the extended hydrophobic group, are surrounded by water-containing hydrated hydrophilic groups. This structure minimizes the interfacial energy and leaves the maximum number of water molecules in mutual contact. Such a structure is called a micelle.

A micelle effectively selects and traps appropriate chemical additives to provide emulsion stability as well as impart desired metalworking lubricant characteristics. It is the uniqueness of the ionic charge and distribution, as well as geometry of the micelle that, when matched to chemical additives, provides the desired emulsion characteristics.

Thus, the bulk properties, as reflected by the micelle structure of an emulsion, and the interfacial properties that exist between two interfaces (the interface between the continuous and discontinuous phases of an emulsion and the interface between the lubricant film and the workpiece) are influenced by the chemistry of the emulsion as well as the chemistry (composition) of the tooling and workpiece.

Surfactants are not alone in their ability to stabilize emulsions. Finely divided solids may also be effective in this regard. Sulfates of many metals, such as copper and aluminum, are efficient stabilizing and emulsification agents for the formation of oil and water emulsions. Similarly, basic metal salts, powdered silica, carbon black, and bentonite have been investigated and determined to also stabilize emulsions.

The effectiveness of these solids is critically dependent upon the method of manufacture of the solid and the resulting particle size, chemical characteristic, structure, and shape. Furthermore, the method of mixing employed to add the solid to the liquid phase can determine the effectiveness of a particular solid additive on an emulsion system.

Emulsion Stability

The breaking of emulsions is generally observable as a splitting of the liquid lubricant into two distinct and separate phases, one consisting of the continuous and the other of the discontinuous phase. However, another common attirbute of emulsions is often observed, called creaming. This separately appearing phase, normally floating on the surface of the emulsion, is not so much a result of the breaking of the emulsion, but rather it is a separation of the emulsion into two emulsions, the original one and the cream, which tends to be richer in the dispersed phase, thus decreasing the concentration of the dispersed phase in the original emulsion.

The cream has a larger particle size and its presence may indicate that a process of de-emulsification (breaking of the emulsion) is about to begin. On the other hand, such mixtures of emulsion particle sizes may be stable enough in metalworking lubricants for a particular application and may provide lubrication superior to fluids with only the smaller emulsion particles.

De-emulsification is the result of the coagulation of the dispered phase. This coagulation is a two-stage process. In the first stage, flocculation takes place; droplets of the dispersed phase aggregate to form larger droplets. As a result, an increase in viscosity may be observed. This aggregated state may be desirable for some metal forming operations. When the aggregated droplets coalesce to form larger and larger droplets, de-emulsification may then result (Fig. 6.5).

When breaking an emulsion is desirable, as in the case of disposal of a spent metalworking emulsion, salts and/or acids are commonly added, thus altering the surface tension at the interface between the dispersed and continuous phases and permitting flocculation and coalescence to take place.

The effect of temperature on stability is not well understood. However, it has been shown that destabilization of the emulsion may occur when high temperatures exist during the metalworking operation.

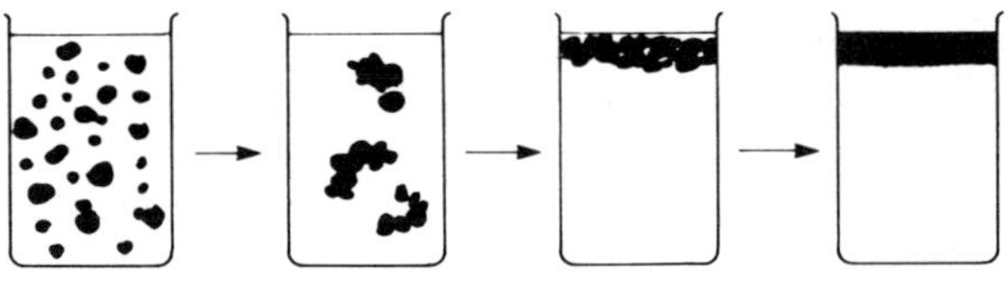

Figure 6.5 Instability in emulsions. (J. A. Schey, 1983. *Tribology in Metalworking.* American Society for Metals, Metals Park, Ohio, p. 154.)

Although the chemical emulsifying system is the most important single determinant of the type of emulsion formed, that is, whether it is a direct emulsion of oil-in-water, an invert emulsion of water-in-oil, or a microemulsion, the emulsifying system developed is also influenced by the method of mixing, temperature, and time.

The oil-in-water emulsion is by far the most widely used emulsion system for metalworking lubrication. The water-in-oil emulsions have found favor as hydraulic fluids and rust preventatives and are infrequently used for metalworking fluids. Microemulsions, referring to the very small size of particle produced, that is, less than 0.1 μm, result in transparent emulsions that have been used primarily as cutting fluids.

Emulsifier Selection

As just described, the emulsifier molecule consists of a hydrophobic group joined directly or indirectly to a hydrophilic group. It is the balance between these two groups that determines the emulsifying characteristics of the molecule. The most commonly used method for evaluating the suitability of a given molecule for producing the desired emulsification characteristics has been designated the hydrophilic-lipophilic balance (HLB) method. An HLB number is determined for each emulsifier and is then assigned so that it can be rated for purposes of application (Table 6.2). Generally, this number is a measure of the water solubility of the emulsifier.

HLB numbers in the range 4 to 6 reflect those emulsifiers selected for water-in-oil emulsions; HLB numbers in the range 8 to 18 may be suitable for the preparation of oil-in-water emulsions. The system has proven to be useful in investigating emulsion types and combinations. However, in many systems subtle binding and size effects complicate the emulsion choice. Further laboratory evaluation is required before the appropriate combination of emulsifier and chemical composition can be selected.

Table 6.2 HLB Values Required for Emulsification

Oily phase	O/W emulsion	W/O emulsion
Acid (oleic, stearic)	17	–
Alcohol		
Stearyl	15	7
Lauryl	14	–
Fatty oil		
Castor	14	–
Cottonseed, tallow	6	–
Palm, rapeseed	7	–
Lanolin (anhydrous)	12	8
Butyl stearate	11	–
Wax		
Beeswax	9	5
Carnauba	15	–
Paraffin, microcrystalline	10	4
Mineral oil		
Paraffinic	10	4
Aromatic	12	4
Kerosene	12	6
Chlorinated paraffin	12-14	–
Tricresyl phosphate	17	–
Methyl phenyl silicone	7	–

Source: After W. C. Griffin, 1979. In *Kirk-Othmer Encyclopedia of Chemical Technology,* 3d ed., Wiley, New York, 1978 vol. 8, pp. 900–930.)

Emulsifiers are classified as anionic, cationic, nonionic, and amphoteric. Petroleum sulfonates are the most common form of anionic emulsifier and, because of their low cost, are widely used. However, when emulsions with this emulsifier are used in hard water, lime soaps result that precipitate and can deleteriously affect the lubricating qualities of the emulsion. Furthermore, they may result in breaking the emulsion. This effect is called calcium sensitivity.

Nonionic emulsifiers have been growing in acceptance because they are not affected by changes in pH or hardness of the water. Typically this class of emulsifiers is formed by the reaction of a hydrophobic hydroxyl-containing compound, namely, an alcohol or phenol with ethylene oxide or propylene oxide. This is a polymerization reaction. The number of oxide groups added can be varied and, thus, greatly affect the solubility and surface-active properties of the resulting polymer. Thus, the emulsifier can be selected to match the chemistry of the system being emulsified.

Forming Emulsions

The temperature during formation of an emulsion, the order of addition of concentrate to the water, and the mixing methods all influence the character of the emulsion, particularly with respect to particle size, as well as its stability.

The temperature of the water used in making up an emulsion should preferably be kept between 50 and 80°F (10 and 27°C). Too high a temperature may deleteriously degrade the components of the concentrate, and too low a temperature may cause difficulty in forming a stable emulsion. Rates of reaction may be so slowed down that the required surfactant films form, if at all, at unacceptably low rates.

In the case of oil-in-water emulsions, to facilitate formation of the emulsion, the concentrate or oil phase should be added to the water. If the reverse order of addition is followed, an invert emulsion may form locally, causing difficulty in reinverting it to form the desired oil-in-water emulsion. The rate of addition of the concentrate, or the discontinuous phase to the continuous phase, may also influence emulsion characteristics and stability. Therefore, slow rates of addition and low volumes at the slow rates of the discontinuous phase should be employed to facilitate the desired formation of the emulsion.

Formulation of metalworking lubricants takes into account the relatively simple mixing facilities available on the plant floor. Thus, from a practical standpoint in production, if reasonable temperatures are maintained, if clean mixing containers are used, and if reasonable mixing of the phases is carried out using small additions of the discontinuous phase to the continuous phase, then an appropriate emulsion should be obtained. In the plant of the supplier of lubricant concentrates, on the other hand, the mixing equipment may be sophisticated and can drastically influence the character of emulsions resulting from the mixed concentrate.

6.1.3 Solutions

A metalworking solution consists of organic or inorganic molecules that are together and separately soluble in water. Frequently, these solutions have been misnamed synthetic fluids. Generally, such solutions do not contain mineral oils but frequently do contain water-soluble polymers, phosphates, amines, sulfur- and chlorine-containing compounds, and possibly surfactants. An emulsion consisting of a dispersion of one liquid phase in another is clearly different from a solution that consists of but a single phase.

Molecules in a liquid are in intimate contact with each other so that the properties of each substance in the solution are influenced by the properties of all others in the solution. These effects may be so energetic that actual reaction takes place in the solution, and the molecules present are quite different from those of the initial substances.

Metalworking solutions are primarily solutions that consist of water as the solvent and numerous organic or possibly inorganic substances present as solute or solutes. The concentration of the solute is generally small relative to the concentration of the solvent. The solubility of substances in water depends upon many factors, including the degree of miscibility or solubility of the particular molecular species present. In the case of solids, it depends upon the relative rates of solution and deposition of the solid in the liquid, as well the possibility of partial miscibility of one molecular species in another. Many pairs of liquids are completely miscible, one in the other; that is, they dissolve in each other in all proportions.

In water systems solution takes place in two principal ways: hydration of ions and hydrogen bonding or protonic bridging. Hydration is the principal process of solution by which ionic substances are dissolved in water. This solubility of strongly bonded molecules, such as sodium nitrite in water, which exhibits strong cohesive forces, is believed to be due to the polar nature of the water molecule. This polar nature derives from the positive charge associated with the hydrogen atom and the negative charge associated with the oxygen atoms in the symmetrically arranged water molecule. The positive ion of the sodium nitrite, namely, sodium, is attracted to the negatively charged oxygen of the water molecule, and the negatively charged ion of sodium nitrite, namely, the nitrite ion, is attracted to the positively charged hydrogen ion of the water molecule. In this way the ionic molecule becomes associated with the polar water molecule and becomes soluble.

The hydrogen bonding, or protonic bridging, comes into play when organic molecules consisting of a carbon chain (which tends to be hydrophobic) and double-bonded alcohol or acid (which tend to be hydrophilic) are in contact with water. If the carbon chain is below a critical length and is associated with a double-bonded structure, such as an alcohol or acid, the organic molecule will be miscible in water through protonic bridging. That is, the OH groups of the organic alcohol, for example, exhibit the same bridging attractive forces as those in the OH group of the water molecule; thus, mutual miscibility occurs.

It is the relationship between attractive and repelling forces in the molecular species in solution that governs the solubility of solutes in solvents. Recently, the hydrogen bonding relationships between water and organic alcohols, polyols, acids, and esters have been used to develop metalworking lubricants with the water as solvent and various water-soluble organic molecules as solutes.

Solutions are relatively insensitive to the effects of pressure. This is not the case with respect to temperature. As temperature increases, solubility of both solids in liquids and liquids in liquids may increase or decrease. Solutes that are completely soluble at room temperature may precipitate at higher temperatures. This inverse solubility with temperature is a phenomenon that has been taken advantage of recently in some metalworking fluids. Glycol esters and glycols, for example, precipitate at elevated temperatures and redissolve when the solution cools. Thus, in a metal cutting or forming operation a solid may deposit and serve as a film during deformation and yet redissolve for further application when the temperature of the fluid is reduced.

6.2 OIL-BASED SYSTEMS

The most ancient and yet most widely used chemical system of lubrication is based on the application of natural oils. The oils to be considered are mineral, animal, vegetable, fish, and synthetic. In this section, we will discuss all these oils and their chemistry, except synthetic oils, which are discussed in the section on synthetic lubricants.

6.2.1 Mineral Oils

Mineral or petroleum oils have formed over long geologic times in the earth's crust as a result of high temperature and pressure on organic matter. The present petroleum industry can be traced back to a day in August 1859, when pipe was driven through loose sand and shale to a depth of approximately 70 ft (21 m), resulting in the first flowing well of crude oil, or petroleum.

Crude oils are complex mixtures containing many individual hydrocarbons or compounds of hydrogen and carbon. These range from methane, the main constituent of natural gas with 1 carbon atom, to compounds containing 50 or more carbon atoms. Crude oils also contain varying amounts of compounds of sulfur, nitrogen, and oxygen, various metals, such as nickel and vanadium, and some entrained water containing dissolved salts. Crude oils range from paraffinic, which are high in paraffin hydrocarbons, through the intermediate or mixed structural oils, to the naphthenic crude oils, which are high in hydrocarbons containing ring structures. The asphalt content varies in different types of crude oils.

On a volume basis, nearly all the world's mineral lubricating oil is derived by refining fractions obtained directly from crude oil. Lubricating oils are made from the more viscous portion of the crude oil barrel, after removal by distillation of the gas and lighter fractions. The refining and processing of lubricating base stock to produce lubricating oils is a complex chemical and chemical engineering problem that will not be treated here.

For this discussion it is sufficient to appreciate that the two principal mineral base oils used to blend metalworking lubricants are either predominantly naphthenic or paraffinic in nature. Naphthenic base stock contains essentially no wax and has a relatively high aromatic oil content. This base stock is more reactive and more sensitive to high temperatures, resulting in oxidation, degradation, and formation of oil-insoluble sludge.

Paraffinic base stocks, on the other hand, must be treated to remove the wax content. They have a relatively low aromatic molecular structure concentration and tend to be less reactive and more stable at high temperatures. The naphthenic oils are often the lubricant base stocks of choice for metalworking lubrication, when additive response and solubility are to be maximized and when temperatures are low or use of the oil is nonrepetitive. On the other hand, paraffinic base stock, which is much more readily available, can, by appropriate selection of the additives, be made into effective metalworking lubricants.

The naphthenic base oil is characterized by a hydrocarbon that has saturated carbon atoms in a ring structure with the general formula C_nH_{2n}. The basic naphthenic structure for a six-carbon ring is one that has single bonds between carbon atoms, or it may be a combination of six- and five-carbon rings (Fig. 6.6). Paraffinic base oils, on the other hand, are characterized by saturated straight (normal) or branched isocarbon chains.

The generalized paraffinic molecule can be symbolized by the formula C_nH_{2n+2}. The basic structure, as shown in Fig. 6.6, consists of a series of carbon atoms with single bonds between them, which shows branched paraffinic chains also with single carbon bonds. Both these rather typical hydrocarbon configurations exist together in both naphthenic and paraffinic base oils. It is the proportion of each that differs. Furthermore, as indicated, in naphthenic base oils there is an appreciable aromatic content. The aromaticity may be reflected in Fig. 6.6, which shows a benzene ring consisting of six carbon atoms but with double-bonded sites.

The viscosity of lubricating oil base stock is determined by the viscosity of the distillate. The concentration of complex molecules increases with vis-

Paraffinic

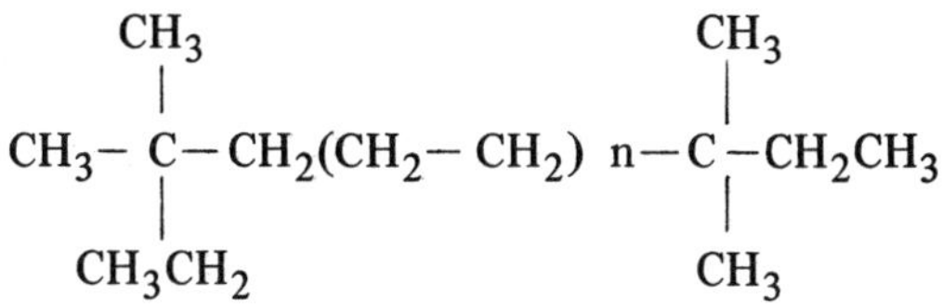

Naphthenic

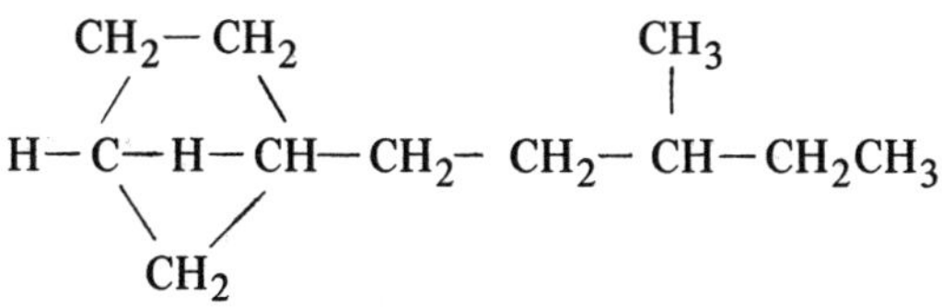

Aromatic

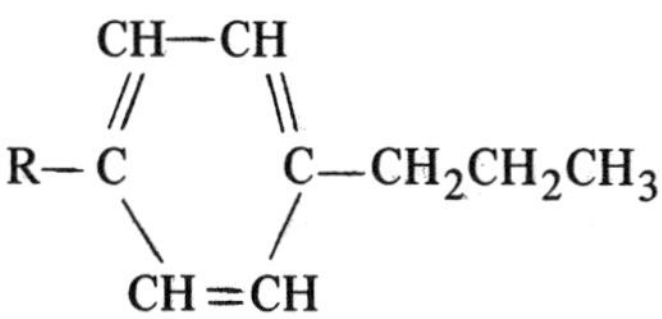

Alkylated Aromatic

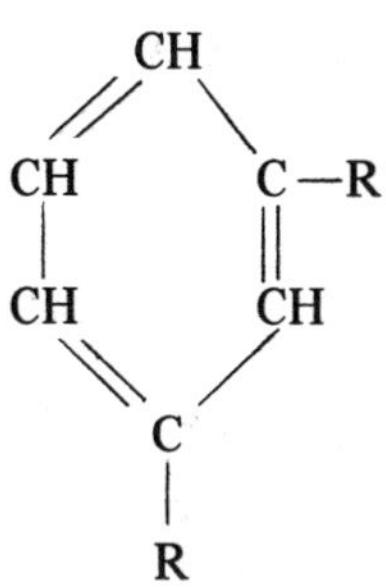

Figure 6.6 Molecular structure of various fluids. (Continued on next page.)

Polybutene

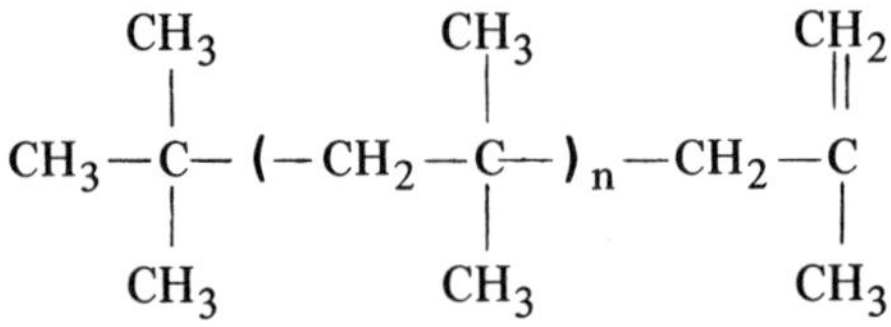

Ethylene Glycol

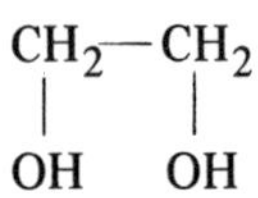

Polyethylene Glycol Ether

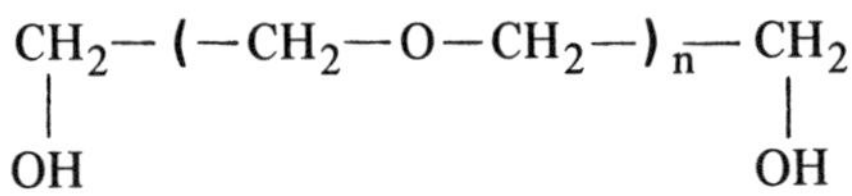

Figure 6.6 (Continued)

cosity; at the same time, the concentration of relatively short chain paraffinic components, as well as simple ring compounds, decreases.

There are advantages as well as disadvantages in considering the use of mineral oil lubricants as compared with water-based lubricants. The mineral oil lubricants are generally stable with respect to biological attack; that is, there are no bacterial or fungal problems. Frequently, there is better lubricity; that is, the lubricating film adheres strongly under high applied pressures. Recycling may be less costly since no separation of oil and water is required; that is, emulsion cracking is not necessary. On the other hand, mineral oils for a given application are generally more expensive. They provide a much lower level of cooling capability. Furthermore, there is the danger of oil contamination in the air, which may lead to health as well as fire hazards.

During the fractional distillation of crude oil a lighter fraction, called hydrocarbon solvents, is produced. Napthas have been used extensively as the principal components of metalworking lubricants. Mineral spirits and low-odor mineral spirits with flash points in excess of 105°F (40°C) are naptha fractions and are often used as metalworking lubricants. These hydrocarbon fractions are used when evaporation of the lubricant is desirable.

Chlorinated solvents are used as fluid lubricants when evaporation of the film is important and there must be no flash point. Furthermore, in some metalworking operations, the chlorinated molecule acts as an EP (extreme-pressure) additive.

In all applications involving the use of solvents as lubricants, care must be taken not to exceed allowable concentrations of evaporating solvents in the air circulating in the workplace.

We will now turn to the consideration of oils derived from vegetable, animal, or fish. All these oils have special characteristics that make them useful for certain metalworking applications.

6.2.2 Vegetable Oils

Rapeseed is used primarily to lubricate the mold during continuous casting of steel. Crambe oil, which is richer in erucic acid than rapeseed oil, has superior lubricity when used as the mold lubricant in continuous casting. Palm oil has been used in large volume for rolling, particularly in cold rolling of strip.

Jojoba oil (a mixture of wax esters) may become a logical replacement for sperm oil in metal cutting fluids. Jojoba oil consists of straight-chain wax esters made mostly of C40 to C42 fatty acids and fatty alcohols (30 and 50%, respectively). In each of the fatty acids and fatty alcohols there is one point of unsaturation with hydrogen atoms, mostly in the cis configuration. Interestingly, this vegetable wax is not a fat or triglyceride but a liquid wax of high moleculear weight. The crude oil contains no fats and little besides the liquid wax and requires no refining for use in most industrial applications. It has a high viscosity index and high flash and fire points. It takes up larger amounts of sulfur than sperm whale oil. It does not darken to the same extent as other oils on sulfurization, and the highly sulfurized jojoba oil remains liquid, whereas highly sulfurized sperm oil requires the addition of mineral oils to do so.

One of the most important properties of jojoba wax is that it is undamaged by repeated heating to high temperatures, and its viscosity does not change with temperature. Furthermore, the wax or oil is not susceptible to oxidation; in other words, it will not become rancid and can be stored for long periods of time.

Rice bran oil, which has a very high free fatty acid content, may become extremely rancid. Rice bran wax may prove to be a suitable material for some heavy-duty presswork applications and may replace carnauba wax, which is used at present.

Castor oil, also known as ricinus oil, is unique among all fats and oils. It is the only source of an 18-carbon hydroxylated fatty acid with one double bond. It is one of the few sources of naturally occurring glycerides that approaches a pure compound. It is a triglyceride (ester) of fatty acids. Approximately 90% of the fatty acid content is ricinoleic acid, an 18-carbon acid with a double bond in the 9-10 position and a hydroxyl group on carbon number 12.

This combination of hydroxyl group and unsaturation exists only in castor oil. The presence of the hydroxyl groups results in relatively high viscosity and specific gravity. Many chemical reactions are used commercially to produce important castor derivatives by acetylation, alkoxylation, amidation, dehydration, epoxidation, esterification, halogenation, saponification, polymerization, oxidation, and sulfation.

The soaps, amine compounds, and esters are used in cutting oils and many industrial lubricants as emulsifiers in soluble oil systems and as sheet-metal working fluid additives. The castor esters have been shown to be useful as cutting oils or components of cutting oils, as well as oil-water emulsifiers for metalworking lubricants. As indicated, the animal, vegetable, and fish oils all contain glycerides of fatty acids. The glycerides are neutral substances, whereas the fatty acids are acid in nature.

Most vegetable oils occur in the seeds or fruits of plants and are obtained by pressing or by extracting with solvents. Since each of the oil-yielding plants produces its characteristic oil, there are many varieties of so-called fixed oils. It is important that all oils used for lubrication should be as free of fatty acids as possible. All fixed oils contain a certain amount of fatty acids, and if the oil is allowed to stand in the air the amount increases and the oil may become rancid. Oils that contain more than 4 to 5% fatty acid should be used with discretion. The fatty acids not only attack and corrode the metal but, at the same time, form soaps that dissolve in the oil and cause it to become thick and gummy. Thickening, however, may be desirable in heavy-duty metalworking applications, such as broaching and forging.

Rosin oil is the principal product of the destructive distillation of common rosin. Rosin oil has a distinctive odor that is not unpleasant, and has been used particularly as a component of grinding fluids. Tall oil is a by-product of the paper pulp industry and is relatively low in cost. It is composed principally of fatty acids and rosin acids. It has been used extensively as a component of cutting fluids.

6.2.3 Animal Oils

Oils derived from animal rendering are infrequently used as lubricants by themselves. They are frequently components of metalworking lubricants.

Lard oil is a secondary product from the manufacture of lard and may be regarded as the liquid part of the latter. It is obtained by allowing lard to crystallize at a carefully regulated temperature and then separating the liquid and solid portions by straining under pressure. Grades of lesser purity are used for lubricating oils. The amount of fatty acid present in different samples also varies greatly. The concentration of fatty acids, as well as the color, vary depending upon the temperature of crystallization used in manufacture.

Generally, the oil is used more as an additive than as a base for metalworking lubricants.

Tallow, which is obtained primarily from cattle or sheep, consists of a mixture of fats. In a good grade of tallow, fatty acids should be under 0.5%. Tallow oil is the liquid part of tallow and is made by melting tallow and keeping it warm for several hours. Much of the stearine content crystallizes and is deposited, under hydraulic pressure, in canvas bags. The principal use of the oil today is as a component of an additive package for metalworking.

Degras, wool fat, and lanolin are fats obtained by washing or extracting the natural fat from sheep wool. They oxidize very little and, therefore, do not become rancid.

6.2.4 Marine Oils

Tuna, herring, sardine, and menhaden have been commonly used as a source of long-chain (18) fatty acids and are highly unsaturated; hence, they are often hydrogenated or sulfurized before use. As components of metalworking fluids they provide good corrosion protection characteristics. Seal and whale oil have been commonly used either alone or in combination with mineral oils for cutting fluids.

6.3 SYNTHETIC LUBRICANTS

Within the context of this book, synthetic metalworking lubricants are fluids that have a definite structure (molecular) and are not naturally occurring. This control of structure results in products with controlled physical as well as mechanical properties. These materials are unlike naturally occurring oils, which, because they are a complex mixture of naturally occurring molecular species, have properties that are an average of the properties of the many species present.

The synthetic fluids more specific in structure are derived from two classes of raw materials, namely, those derived from petroleum (polyalphaolefins and alkylated aromatics) and polybutenes. At present, synthesized hydrocarbons are most widely used. Nevertheless, growing in importance are synthetic lubricants of a second type based on nonpetroleum sources, particularly derivatives from the fractionation of vegetable oils. The important classes are organic esters and polyglycols (Fig. 6.6).

Before continuing the discussion of the chemistry of synthetic fluids, the nomenclature problem very common in the industry, among the producers as well as users of metalworking lubricants, should be clarified. It is common practice to describe solutions of various kinds containing different molecular species to provide lubricity, particularly water-based solutions, as synthetics. This nomenclature has been developed because in almost all cases water-based solutions have replaced either soluble oils containing oil or straight oils. The contrast has led to the adoption of the word synthetic to describe non-oil-containing lubricants. In this book we have attempted to clarify the nomenclature and to use the chemically acceptable terminology, thus reserving the word synthetic for those synthetic fluids suitable for metalworking application, either alone or in combination with appropriate additives.

Polyalphaolefins are comparable to a highly refined paraffinic mineral oil with excellent shear stability. They have been successfully used with EP additives for metalworking applications.

Alkylated aromatics and, specifically, di-alkylated benzene, have good extreme-pressure lubricating properties. The alkyl group attached to the benzene ring has a normal paraffinic structure (Fig. 6.6). Such operations as swaging or tube expanding with internal mandrels have been successfully performed with lubricants formulated with this base stock.

Polybutene is a polymerized material consisting of butenes and isobutenes. The lower molecular weight polymers have useful lubricating characteristics. Furthermore, if parts coated with this lubricant are heated for stress relieving or other metallurgical purposes, the polybutenes decompose completely, leaving no deleterious residue.

Among non-petroleum-based synthetic fluids, dibasic acid esters have good lubricating properties and, furthermore, have the ability to suspend small-particle contaminants of the fluid, such as oxides, thus helping to maintain clean metal surfaces. Commonly used acids vary in carbon chain length from 4 to 8. Adipic, azelaic, and sebacic acids are commonly used. Tridecyl and trimethyl alcohol are common lubricant components.

Polyol esters are formed by reaction of a monobasic acid with a polyhydric alcohol. A broad range of properties can be achieved by varying the alcohol and acid proportions. The acids are usually obtained from animal or vegetable oils and contain from 5 to 10 carbon atoms. These esters have good lubricity with good film strength at elevated temperatures.

Polypropylene or polyethylene glycols may be polymerized to form both water-soluble and water-insoluble types (Fig. 6.6). If heated to their decomposition temperature, the water-soluble lubricants, that is, the polyglycols of lower molecular weight, completely volatilize, leaving no deposits. Furthermore, these chemicals are used as a base fluid in which other water-soluble additives may dissolve. Water glycol solutions have many other applications. They have been successfully used in metalworking, particularly in the lubrication of rolling, drawing, and machining of aluminum.

Phosphate esters, which maintain their integrity at relatively high temperatures and exhibit good wetting characteristics, have been used as additives in mineral oil to reduce friction. A typical phosphate ester is tricresyl phosphate, with three-ring structures attached to the phosphorus atom through an oxygen atom.

In general, all the synthetic fluids, because of cost, have restricted application as lubricants. They have, instead, generally found application as additives or components of lubricants using other base stocks, either water or mineral oil.

6.4 SUSPENSIONS, PASTES, SOLID LUBRICANTS, AND COATINGS

In this section lubricating systems critical to the economical operation of metalworking processes, such as wire drawing, hot forging, hot extrusion, and sheet-metal drawing, are described. These systems, in large measure, rely on the mechanical and physical properties of the lubricant to provide the necessary film characteristics. In some processes the properties of the lubricating film are indispensable for economic metalworking. For example, the development of glass coatings in France in the early 1940s made hot extrusion economically feasible. The development of phosphate coatings in Germany in the 1930s made cold forging of steel an economical process for producing parts.

In this section lubricants in the form of suspensions, pastes, solid lubricants, and coatings are described. Examples are chosen to illustrate the relationship between lubricant and the metalworking process. As has been the case previously in this text, there is an overlap of lubricants between several states (graphite is both a solid lubricant and a suspension). The examples chosen, however, should clarify any difficulty in understanding the principles involved.

6.4.1 Suspensions

Suspensions in metalworking fluids consist of fine particles of a particular solid lubricant dispersed throughout a carrier fluid. The particles normally vary in size from 1 to 100 μm. Often, the particle is totally insoluble in the carrier medium; nevertheless, its surface is wetted by the carrier.

A dispersion may be considered as resulting from the occurrence of three sequential processes:

1. Wetting of the particle by the carrier fluid
2. Reduction in size of particle clusters
3. Stabilization of particle suspension

The effectiveness with which each of these processes is carried out depends upon the chemistry of the particle and the carrier fluid, the surface interfacial atomic forces, and the effectiveness of mixing.

Coagulation or flocculation (these terms are used here interchangeably) must be prevented if the suspension is to remain stable over time. This involves creation of a repulsive force between adjacent particles to prevent their coagulation. Normally, this charge is provided by appropriate chemically adsorbed layers added to the dispersed systems. Surface-active agents are important in these systems. They often are specific to a particular system and may not be used indiscriminately when considering one dispersion system relative to another. Each case should be considered on its own merits.

Surfactants may promote wetting of the solid by the liquid. However desirable this ability, of even greater importance is the effect on adsorption energy barriers. Surfactants, which are dispersing agents, must have the ability to form energy barriers to aggregation. Some examples of dispersing agents used in aqueous systems are lignin sulfonates, acrylic acid-based copolymers, and copolymers of propylene oxide and ethylene oxide. Suspension of solid particles in nonaqueous fluids, such as mineral oil, have been effected through the use of alkyl benzenes and long-chain amines.

With this brief description of the chemistry of suspensions as a foundation, specific metalworking lubrication systems will now be considered. Warm and hot forging of steel and other metals have relied on the use of graphite suspended in mineral oil and, with increasing frequency, water. The lamellar structure of graphite, its stability at relatively high temperatures (about 1000°F; 540°C), and the beneficial effect of air and water vapor on its lubricity, all have encouraged the application of graphite in a carrier as the most cost-effective high-temperature lubricant. MoS_2, suspended in mineral oil, has been used extensively in wire drawing, where pressure is high during deformation but temperatures are below 600°F (315°C), and hot forging.

6.4.2 Pastes

Heavy-duty pressworking and bar drawing are operations that have been successfully carried out using pastes as lubricants. These pastes may be water or oil based and pigmented or nonpigmented, and vary in viscosity from one that is readily pourable to a paste that is almost a solid.

The structure of pastes depends upon the manufacturing procedures used, as well as the concentration of the component present as the thickener.

Thickeners may be water- or oil-soluble polymers, treated naturally occurring clays, and soaps and/or surfactants.

Polyvinyl alcohol and carboxymethyl cellulose are two commonly used water-soluble polymers. These polymers dissolve completely in water to yield clear dispersions of a highly swollen gel that contains several thousand times their weight in bound water. They are efficient thickeners at low concentrations, have good gel strength, and maintain viscosity at higher temperatures.

Nonpigmented sodium or potassium soaps are widely used in pastes. These pastes, consisting of soaps based on vegetable or animal fats and water, have very good lubricating characteristics. The pastes have the appearance of mayonnaise and generally consist of anywhere from 30 to 60% fat. They can be diluted with water at ratios of 1:1 to 1:5 but are not recommended for circulating systems because of excessive foaming.

Pigmented soap compounds are used for more severe sheet drawing or wire drawing operations. Examples of solids that may be used are graphite, molybdenum disulfide, sodium carbonate, and mica.

Pastes illustrate a special state of aggregation that can be achieved to provide desirable viscosity characteristics for many metalworking operations. Their rheological properties can be complex. Their viscosity may increase or decrease as the temperature and pressure increase, depending upon the particular chemical system involved.

The most serious disadvantage in the application of pastes to metalworking operations is the buildup that occurs on the dies during metal forming. This buildup of solid films can interfere with the forming operation by altering the dimensions of the dies. As a result, parts can crack, definition of detail can be lost, and ongoing cleanup costs can be excessive. Furthermore, the properties of paste may change with time due to the splitting of the system or thickening to a solid. Foaming is also a problem associated with the use of soap-based systems.

6.4.3 Solid Lubricants

Several solid lubricant systems are important in metalworking operations, principally in those operations in which very high pressures or temperatures result from the deformation process. The thickness and film strength requirements for adequate lubrication in wire drawing of high-temperature alloys or hot extrusion are most frequently satisfied by the solid lubricants.

Solid lubricants generally consist of

1. Soaps and soap combinations
2. Graphite and molybdenum disulfide, alone or in combinations

3. Glasses
4. Metals

A wide range of solid materials is infrequently used or under development:

1. Graphite + boric oxide
2. Phthalocyanine + graphite
3. Magnesium metaborate-graphite mixture in water
4. Boron nitride
5. Nylon and polytetrafluoroethylene

Aluminum stearate, calcium stearate, and other metallic soaps are often used alone or in combination with molybdenum disulfide and/or graphite as effective wire drawing lubricants. These soaps are relatively inexpensive and provide tough adherent films, permitting large deformations.

Graphite at temperatures between 85 and 1000°F (30 and 540°C) is a good lubricant and has been successfully used in the hot extrusion of steel tubes as well as in the drawing of tungsten and molybdenum wire at 930 to 1300°F (550 to 700°C). Graphite mixed with some of the metal oxides (tin oxide and copper oxide) has been used as a lubricant in the extrusion of steel at 2100°F (1150°C). The relatively easy cleavage of the lamella structure of graphite is responsible for its lubricating characteristics. The presence of oil and/or water with graphite appears to improve its lubricating characteristics and is often utilized in developing a paste or suspension for appropriate application of the graphite (Fig. 6.7). Similar lamellar flow characteristics of molybdenum disulfide account for its excellent lubricating properties (Fig. 6.7). However, molybdenum disulfide is not an effective solid lubricant above about 930°F (500°C), due to oxidation. In combination with other solids it is used as an effective lubricant up to 1800°F (1000°C).

The initial development of glass as a lubricant for hot extrusion has been followed by an expansion of the chemistry of glass lubricants to provide a broad range of temperatures and film characteristics desirable for many hot-working applications. The amorphous film of molten glass has excellent wetting properties of the metallic surface and provides a long-lasting film that becomes fluid during the forming operation due to temperature rise.

Glasses are made for use from 850 to 4000°F (450 to 2200°C). Generally, their success depends upon an unusual combination of properties: they soften over a range of temperatures, gradually changing viscosity with temperature, are quite stable over this wide temperature range, and do not chemically attack the metals being formed. Furthermore, they have low thermal conductivity.

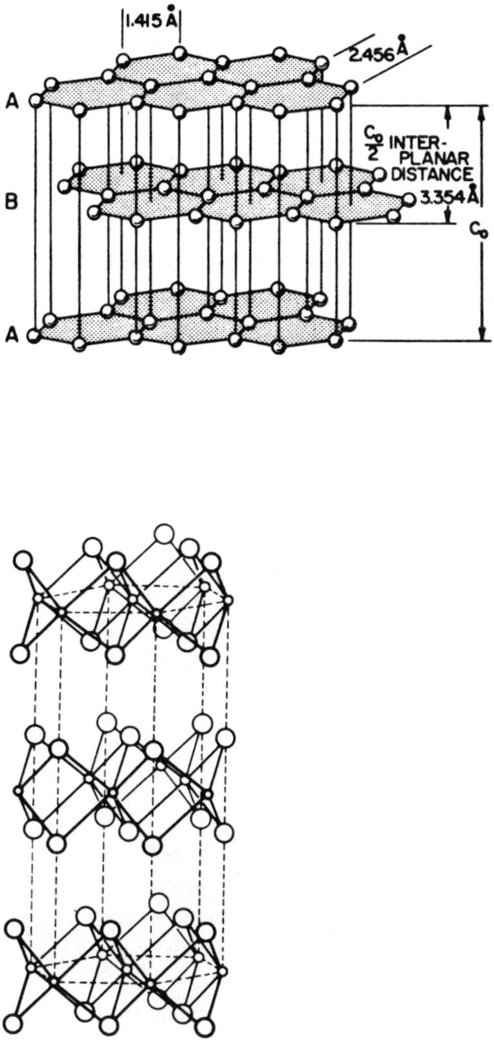

(b)

Figure 6.7 (a) Structure of graphite. (b) Structure of molybdenum disulfide.

For service above 2000°F (1100°C), glasses are usually made of silicates. For intermediate temperatures they are generally borates, and for low temperatures they are normally phosphates. Alkali phosphate-based lubricants, which melt above approximately 750°F (400°C), form adherent coatings with good wetting capability.

Metals have been used successfully in volume production of wire as well as for canned rolling and extrusion operations, for example, copper coating of a billet prior to reduction to plate when the material is difficult to deform at rolling temperatures. In severe heading operations, copper has been successfully used for specialized metalworking applications, such as drawing stainless steel rod and wire.

6.4.4 Coatings

Metals that are difficult to form or that undergo severe bulk deformation, thus making extraordinary demands on the lubricant, may be coated prior to working. These coatings, applied either at the mill or at the fabricating shop, may be used alone as the sole lubricant during forming or may provide an undercoating with a liquid lubricant applied over it during the deformation process. It is common practice to lime coat or borax coat steel bars and tubing prior to a drawing operation. Steel slugs are often phosphate coated prior to cold extrusion, and stainless steel is frequently given an oxalate coat prior to bar or tube drawing. These coatings improve the operation through their presence as a durable film that prevents metal-to-tool contact. A liquid lubricant with necessary lubricity and penetration characteristics is almost always applied during processing to maximize the effectiveness of lubrication.

Soap in powdered form may consist of water-soluble soaps, such as sodium, water-insoluble soaps, such as calcium stearate, and other additives, such as borax or lime. They are commonly used in wire drawing since the solid soap film provides effective film strength as well as good lubricity. Solid lubricants, such as graphite and/or molybdenum disulfide, may be also be added to further improve the lubricity of the soap coatings.

Water-soluble as well as oil-soluble polymers are used both alone and as suspension media for solid lubricants, such as graphite and molybdenum disulfide. Some of the water-soluble polymers that have been used are polyvinyl alcohol, carboxymethyl cellulose, methyl cellulose and cellulose ethers, polyacrylamides, polyvinylpyrolidone, and polyethylene oxide polymers. Polymer solutions may also contain additives to provide corrosion protection and lubricity, or, in some instances, may be used alone. Most commonly, these coatings are used in sheet-metal forming. The sheet metal is pulled through or dipped in the water solution and is subsequently heated so that the water evaporates, leaving a polymeric coating on the surface.

Other bonded coatings consist of a base material, such as a phenolic resin or a silica organic resin, sodium silicate, or polyisobutylene, with graphite and/or molybdenum disulfide suspended in the carrier. The coating is applied by some appropriate dip or spray application method. It is then heated to promote film formation and adhesion.

Coatings with excellent corrosion protection have been used for many years as slushing oils. Recently, coatings similarly applied at the steel mill have been developed and provide excellent corrosion protection as well as lubricating properties. Cold-rolled as well as galvanized steel sheet has been successfully formed with only this coating. All lubricants previously used at the press have been eliminated, and such parts as floor decks, dashes, and deck lid liners for automobiles have been successfully formed. These oil-like coatings containing EP additives do not stain copper and can be used successfully for many drawing operations.

In general, coatings have economic limitations involving energy, control of deposit geometry and chemistry, and removal. Hence they have had limited applications. On the other hand, some coatings, such as the phosphate coatings applied prior to cold heading, are necessary and provide the only feasible method for carrying out cold-forming operations.

6.5 ADDITIVES

Metalworking processes that require lubrication vary widely in conditions of deformation. In less severe operations, hydrodynamic (thick-film) lubrication predominates; that is, the film thickness at the tool-workpiece interface is sufficient to avoid metallic contact between mating surfaces. This type of lubrication is dependent upon the presence of a film with very large thickness compared with the average size of the lubricant molecule, so that the physical and macroscopic properties (principally viscosity) are significant.

On the other hand, under more severe deformation, the thickness of this film is greatly reduced and, in some cases, such as iron sulfide and iron chloride, can be of the same order of magnitude as the lubricant molecule. This constitutes the boundary lubrication regime wherein the chemical and physicochemical properties of the lubricant, particularly the additive system contained within the lubricant, become of great significance.

As is now understood, an array of lubrication conditions extends from the boundary, at one extreme, to the hydrodynamic at the other, which may or may not occur during a specific metalworking operation. These lubrication regimes are discussed in more detail in Chap. 5. They are reviewed briefly here because of the use of specific additives in metalworking lubricants to influence the mechanical properties of the film as well as its chemical reactivity. Additives are used as a component of the lubricant for many other reasons. Various additives and their chemistry as they influence lubricant properties, as well as their functions, are described below.

Emulsifiers

As just described, anionic and nonionic surfactants are both employed to formulate emulsifiable metalworking fluids. Some common nonionic emulsifiers are ethoxylated alcohols or phenols, fatty acids plus alcohols, monoglycerides, esters of propylene glycols, and esters of sorbatan. Some anionic emulsifiers are carboxylic acid salts, soaps of long-chain fatty acids derived from vegetable oils, abietic acid derived from rosin, and naphthenic acid; and alkane sulfates, sulfated alcohols, alkane sulfonates, phosphate esters and alkyl aromatic sulfonates, and natural sulfonates, a by-product of white oil refining or synthetic sulfonates. The latter are often produced by appropriate sulfonation of a polymer.

Nonionic emulsifiers are attractive because they cover the whole hydrophilic-lipophilic balance scale. In general, nonionic emulsifiers are more expensive, and therefore, anionic emulsifiers are more widely used in lubricant formulation.

Soaps, which are salts of long-chain fatty acids derived from fats and oils, tallow, coconut oil, palm oil, and others, are the most widely used anionic emulsifiers. Sodium or amine soaps are commonly used. However, these products are sensitive to hard water and form hard-water soaps, which can be objectionable since they build up and deposit on both the workpiece and in the lubricant reservoir. The sodium sulfonates, resulting from the acid refining of white oil, have been the economical emulsifiers used for a long time. Their erratic supply in recent years, based on crude oil availability, has led to the development of synthetic alkylaryl sulfonates. If the primary emulsifier is a naturally occurring or synthetic sulfonate, a secondary emulsifier, such as ethoxylated alcohol, may often be used to provide the desired emulsion stability.

It is important to appreciate that these emulsifiers, which have strong film-forming characteristics at the interface between the oil and water phases, also operate at the interface between the metal surface being formed, the tool, and the emulsion. The secondary effect of emulsifiers on corrosion illustrates a phenomenon that is often encountered in various additive systems. The additive has a primary function and a secondary function and often affects other characteristics of the metalworking fluid. This complex interaction is a result not only of the chemical interactions that occur between the various components of the metalworking fluid but is also a response to the temperature and pressure effects inherent in the metalworking operation.

Couplers

Hexylene glycol, butyl carbitol, carbitol, and methyl cellulose are widely used couplers. The performance of an alcohol or glycol depends upon its HLB value, as well as on its structure and molecular weight. Also, as in the case of the primary emulsifiers, these coupling agents can have a significant influence on the corrosion-protection characteristics of the metalworking fluid.

Thickeners

Some understanding of the science of rheology is desirable to discuss the deformation and flow of liquid lubricants. Rheology is the science that deals with deformation and flow of matter. The equation

$$\text{Viscosity} = \frac{\text{shear stress}}{\text{shear rate}} \tag{6.1}$$

describes the relationship of applied load and shear rate to viscosity, where viscosity is a measure of resistance to flow. It is this resistance that is of primary importance in the operating characteristics of metalworking fluids. Viscosity characteristics vary with the applied stress, depending upon the system under study.

Thickeners are added to the lubricant to increase its viscosity under the flow conditions in the particular metalworking process. Newtonian flow is descriptive of a lubricant with viscosity independent of the shear rate. This behavior is typical of solvents and oils. Dilatant flow occurs when viscosity increases with increasing shear rate and is characteristic of some highly pigmented systems.

Pseudoplastic flow is a term used to describe the decrease in viscosity that occurs as the shear rate increases (non-Newtonian). In this use, the structure that exists at one viscosity at a given shear rate breaks down as the shear rate increases and the viscosity drops. As the shear rate decreases, the structure rebuilds and the viscosity increases. In thixotropic flow the viscosity decreases as the shear rate increases. With decreasing shear rate the viscosity recovers slowly due to a temporary loss of the original structure. However, given sufficient time, the original viscosity returns.

Organoclays, that is, clays that have been chemically treated to influence the wetting, shear, and polar activation of the clay, are common thickeners.

Polymers may also be used, but they depend upon swelling, shear rate, and time for the desired increase in viscosity. Combinations of these materials are also used. Natural gums, such as gum arabic or gum tragacanth, as well as casein, gelatin, and albumins, have been used as thickeners. α-Substituted acrylic acid esters, polyvinyl alcohol, and phenolformaldehyde resins are typical of the polymers that may be used as thickeners. Silica and alumina, as well as the hydroxides of many of the heavy metals, are suitable for gelling lubricating oils.

Tackiness Agents

Tackiness agents are added to increase the bonding strength of the lubricant to maintain its integrity under load, as well as to improve the adhesiveness of the lubricating film to the metal surface. The most commonly used additives are organic polybutenes and polyisobutylenes.

Detergents

In many metalworking operations, complex deposits consisting of metal fines, metal oxides, tramp oils, oxidized lubricant components, grease, and other substances may deposit in the system, causing cleanliness problems, providing food for bacteria, and initiating rust. During continued operations, because of the temperature and pressure involved and the catalytic effects of the metal and/or metal oxides present, as well as through biological degradation due to mold and/or bacteria formation, the lubricant that is almost always designed to operate at a pH in excess of 8.5 begins to approach a pH of 7.0. This loss in alkalinity may lead to corrosion and/or staining.

Detergent is a name that has been given to chemical compounds that have the ability to reduce or to prevent deposit formation and, at the same time, neutralize the acid products of lubricant degradation. Along with the development of sulfonates as primary emulsifiers, it was soon realized that these materials also had detergent properties. Long-chain alkyl-substituted benzene sulfonic acids, as well as the standard petroleum sulfonic acids, have been used for this purpose. Sulfonates treated with an excess of the metal salt, such as barium hydroxide, result in "overbased" basic salts. This excess of basic salts is capable of additional neutralization of acid contaminants. Phosphonates and thiophosphonates, phenates, and alkyl-substituted salicylates are also used as detergent additives.

Plasticizers

Some additive systems used in metalworking lubricants produce a film that may have undesirable characteristics. For instance, these films may be sticky and may preferentially collect hard-water soaps when they are formed in emulsions. They may also build up oxide deposits when, for example, aluminum alloys are machined or when steels are rolled. These sticky films can become hard and adherent and may be very difficult to remove. Furthermore, they can mar the surface of the formed part.

High-molecular-weight esters codeposit and "plasticize" or nullify the tackiness or stickiness associated with the original film, thus preventing the conditions just described. Polybutenes appear to behave in a similar manner.

Antioxidants

Because of the operating conditions during metalworking, namely, reactive metal surfaces, and due to the presence of oxygen and elevated temperatures, many of the organic components of the lubricant tend to oxidize, losing their chemical integrity and, in turn, their effectiveness. This effect of oxidation is cumulative, with the degraded products building up over the extended production cycle. Oxidation often leads to a change in the viscosity of oil-based lubricants and may result in the deposition of carbonaceous matter. Furthermore, in all the fluid lubricants of interest, the development of acidic contaminants may occur, along with the deposition of partially degraded components of the lubricant.

A broad array of possible antioxidants may be used in metalworking lubricants to prevent or retard oxidation. Sulfides, phosphites, amines, and zinc dithiophosphate have been used for this purpose. Several specifically designed chemical compounds are also being used successfully as antioxidants. They are phenols, aromatic amines, such as hindered phenols, and *N*-phenyl-β-naphthylamine. Prevention of degradation due to oxidation is important, particularly in metalworking operations that may produce high temperatures.

Antimisting Agents

In many metalworking operations, extreme agitation of the fluid results in airborne contamination of the atmosphere by the fluid. Concentration of these materials in the air is subject to regulations promulgated by OSHA. The misting problem is particularly associated with machining using oil-based lubricants. It has been observed that the addition of very small quantities of polymers, such as acrylates or polybutenes, reduces mist formation. The oil and polymer build together to form particle sizes that are sufficiently large and heavy that they are much less readily airborne.

Oiliness Agents

Particularly in the case of oil- or solvent-based lubricants, an improvement in film strength is often required to reduce the possibility of seizure between the tool and workpiece. Fatty acids, soaps, and organic phosphites and phosphates are commonly used for this purpose.

The films produced are effective under low operating temperatures and pressures but fail under more severe conditions of metal forming or machining. It is difficult to quantify the operating conditions, since they vary depending upon the tool and workpiece materials. Temperatures are generally below 250°F (120°C), and pressures are low.

Dispersants

In metalworking, particularly of nonferrous materials, such as copper and aluminum, surface appearance is frequently important. Agglomerates of contaminants and/or degraded components of the lubricant should not be permitted to deposit on the metal surface and affect the finish or appearance. Polyacrylates have been successfully used as additives that suspend these contaminants or degradation products, thus reducing their tendency to deposit on the metal surface.

Extreme-Pressure Additives

Extreme-pressure additives are additives activated predominantly by the temperatures generated during metalworking. An effective film interposed between the tool and workpiece prevents welding of these surfaces. In the case of extreme-pressure additives, the primary film formation takes place by direct chemical reaction between the additive and the surfaces. These additives are primarily sulfur (sulfurized esters of fatty acids) or chlorine (chlorinated hydrocarbons), reacted organic esters, fats, or oil. Metallic sulfides, chlorides, or soaps are the main constituents of the film that results from the reaction of the additive with the metal surface. Phosphorus additives, such as the esters of phosphorus acid, are also frequently used. Molecules that may have both chlorine and sulfur on the same molecule may show enhanced film-strength formation during metalworking.

On the other hand, some additives act through thermal degradation processes that may be accelerated either by the reactivity of the metal itself or through the presence of oxygen. Metal dithiophosphates and phosphorus-containing organic compounds operate in this way. The reaction films at the surface are very complex, and the final products in the film often have low molecular weights. They may consist of oxides, metal phosphates, or sulfates.

Solid lubricants, such as molybdenum disulfide, are often added and form a film that is adherent to the surface of the metal being formed, thus effectively separating the tool and workpiece, as is required of a good lubricant. Complex borides have been found to act in a similar way. Finally, various complex polymers, which may be oil or water soluble, also may form during in situ polymerization.

The principal extreme-pressure additives are those containing sulfur and/or chlorine. These compounds may be oil or water soluble, depending upon the structure of the molecule containing the sulfur or chlorine. Both the activity, which may lead to staining or corrosion in extreme cases, and the film characteristics are dependent upon the processing conditions and the structure of the molecule containing the sulfur or chlorine. The sulfur- or chlorine-containing film formed is, therefore, responsive to the complex circumstances involving chemistry, metallurgy, geometry, and the temperature and pressure developed during the metalworking operation.

Passivators

Such metals as copper and copper-based alloys, aluminum and aluminum alloys, and zinc-coated steel are all sensitive to staining brought about by atmospheric conditions and through chemical reactions with lubricant contaminants, as well as by the lubricants. Organic diamines, heterocyclic sulfur, and nitrogen compounds have been used successfully as additives to reduce the number of active sites on the metal surface, thus minimizing the potential for staining.

Antifoaming Agents

Foaming in metalworking fluids can be very disruptive. It can mask the operation, interfere with the cleanup of the reservoir, complicate fluid flow, change the coolant activity, and, in some circumstances, require major cleanup operations. Solutions, emulsions, and suspensions all may foam to a greater or lesser extent, depending upon the possible occurrence of the following three disruptive factors: a change in foam bubble size, thinning of the walls of a bubble, and rupture. The foam itself is a two-phase system consisting of a relatively large vapor phase dispersed in a small volume of liquid. The foam bubble has two major shapes: a spherical foam, which tends to be less stable, and a polyhedron-shaped foam, which tends to be more stable.

To destabilize a foam, the surface free energy of the film must be reduced. One approach is to add a antifoaming agent, which neutralizes the chemical potential and the electrical contribution of the film formers producing the bubble. Another approach is to add a surface-active agent that lowers the viscosity at the surface of the bubble. Finally, surface transport rate can be increased to prevent time for bubble wall repair. Polyalkoxysiloxanes, fumed silica, high-molecular-weight amides, and polyglycols are effective antifoaming agents.

Alkaline Reserve Agents

As described previously, contamination, biological attack, chemical interactions, and high temperatures may result in a gradually increasing concentration of acidic constituents in the metalworking fluid. Furthermore, in discussing emulsifiers, the development of "overbased" sulfonates and their ability to neutralize acidic contaminants were also described. Another effective way of nullifying acidic component buildup is by the addition of alkaline reserve agents. Amines, such as tri- and diethanolamine, as well as alkanol amides, are used as effective acidic neutralizing agents.

Solid Lubricants

Graphite, molybdenum disulfide, polytetrafluoroethylene, mica, copper, zinc, and lead powder are among the solid lubricants that may be suspended in metalworking fluids to improve the film strength and its stability. These lubricants generally deform under metalworking pressures and form thin films, often only mechanically adherent to the workpiece or tool surfaces. Nevertheless, the film provides effective separation of the tool and the workpiece.

Odor Mask

Components of the lubricant frequently have undesirable odor characteristics. Sulfur-containing additives are particularly prone to this problem. If an appropriate competing odor mask, that is, a compound that has a distinctive odor of its own, is added to the lubricant, the disagreeable odor perceived by users of the fluid may no longer be detected. Pine oil, cedar oil, and sassafras essence are among the common odorants that have been successfully used to control odor in lubricants.

Corrosion Inhibitors

Corrosion, either assisted or initiated by lubricating fluids, has been a problem in metalworking operations. Particularly troublesome has been corrosion associated with the use of water-based fluids. In oil-based lubricants, there appears to be competition for the surface between EP additives of various types and corrosion inhibitors, such as alkyl-substituted succinic acid. In general, the presence of corrosion inhibitors does not affect the characteristics desired through use of the EP additive. On the other hand, the reverse may not be true, that is, the presence of EP additives in systems containing competing surface-active corrosion inhibitors may reduce the effectiveness of the EP additive.

In water-based systems, on the other hand, electrolytic phenomena take place in which current flow must be prevented. Depending on the metals as well as the fluid used, the tools and/or die surfaces may be anodic. Anodic passivators, such as the chromates and nitrites, have been successfully used to prevent corrosion in such systems. Zinc salts, organic phosphonates, benzotriazoles, sulfonates, borates, and organic carboxylates have found broad use as cathodic inhibitors. Depending upon the system, combinations of cathodic and anodic inhibitors may be more effective then either one alone.

The emulsifier in water-based systems and/or alkaline reserve agents, such as triethanolamine, may serve as corrosion inhibitors. Many other organic amines and metal salts of organic sulfonates provide corrosion protection. Phosphates as well as certain unsaturated fatty acids are also effective corrosion inhibitors. Toluyltriazole is a copper alloy metal corrosion inhibitor. In addition to the specific ionic contribution of the corrosion inhibitor to the system, film-forming additives designed to provide a mechanical barrier when used with ionic passivators may provide still further protection. Various polymer films have also been used in this connection, both water soluble in the case of water-based fluids and oil soluble in the case of oil-based fluids.

Viscosity Index Improvers

In most metalworking operations, to control the mechanics of the operation it is important to maintain as constant a viscosity as possible. In general, it would be ideal if viscosity is independent of temperature. Such a lubricant would have lower internal friction forces during deformation at low temperatures and reduced degradation of the lubricant at high temperatures.

Viscosity index improvers are principally of the following types.

1. Polyisobutenes
2. Polymethacrylates of various chain lengths
3. Vinyl acetate-fumeric acid ester copolymers
4. Polyacrylates, that is, copolymers of various chain length alkyl acrylates
5. Alkylated polystyrenes

Small additions (0.5 to 1.0%) of these chemicals can significantly reduce the variation in viscosity of metalworking fluids as the temperature varies from near 0 to over 400°F (−17 to 200°C).

Dyes

To assist in identification of the metalworking fluid, as well as provide a maker to locate any leakage or spills, dyes are frequently added to the lubricant. Both oil- and water-soluble dyes are available and may be used as appropriate for a particular system.

Friction Modifiers

Some metalworking fluids are treated with friction modifiers to improve flow under load. These materials have low internal friction characteristics and thus reduce power requirements. Complex esters, as well as such compounds as ditertiarynonylpolysulfide, are used for this purpose.

Antimicrobial Agents

Antimicrobial agents are particularly important in water-based lubricants. The microbial concentration in industrial lubricants may range from about 0 to 10^9 microorganisms per milliliter. The microbial problem is primarily associated with the growth of three separate species: anerobic sulfate reducers, such as *Desulfovibrio desulfuricans,* aerobic bacteria (primarily *Pseudomonas* species), and coliforms and imperfect fungi, such as *Fusarium* and *Candida.*

The sulfate-reducing bacteria results through its growth in the well-known "Monday morning stink." These organisms are anaerobic, do not require air, and, hence, grow at the bottom of the sumps. The aerobes, on the other hand, are by far the most prevalent bacteria in metalworking lubricants. They appear to grow and feed on everything and anything in the system and are the primary cause of chemical degradation.

Large fungal populations rarely develop in water-based emulsions. They are more frequently associated with solutions, particularly those containing fatty acid components. In these systems, large fungal populations can develop. The fungi are apparent as large slimy masses, or they may appear as cottonlike balls. They can clog filters, interfere with the application of lubricants, and, in general, cause employee dissatisfaction.

A large group of biocides may be used effectively in metalworking fluids. However, it is important to understand and determine the compatibility of a given additive with a specific metalworking fluid. Furthermore, combinations of additives may neutralize their effects rather than be of benefit. Some well-known biocides used in the United States are

1. Kathon 886MW, a mixture of 5-chloro-2-methyl-4-isothiazolin-3,1 (8.6%) and 2-methyl-4-isothiazolin-3,1 (2.6%)
2. Grotan HD2, a mixture of 2-chloro-*N*-hydroxymethylacetamide (39%), sodium tetraborate (41%), and potassium iodide (39%)
3. Bioban P 1487, consisting of 4-nitrobutylmorpholene (70%)
4. 4-Ethyl-2-nitrotrimethylene dimorpholene (20%)
5. Trisnitro, consisting of tris-hydroxymethylnitromethane.

Certain concentration effects exist of which operators, particularly, should be aware in the use of biocides as additives in metalworking lubricants. The concentration of the additive, unless controlled, can result in an increase in growth of the microbe before it dies. Too low a concentration will not affect microbial growth. The correct amount will reduce microbial growth, a little too much will grow the microbe at an accelerated rate, and too much may result in complaints from the workers and cause dermatitis. The choice of concentration depends upon the chemical system and the biocide used.

Similar observations have been made concerning the effect of hard-water salts (in the water used) on the biocide. Very low concentrations of salt may reduce microbial growth but may result in excess foaming. An optimum concentration of salt in the water, on the order of 100 parts per million, potentiates the effects of the bactericide. Higher concentrations of salt can nullify the effectiveness of the biocide.

Chelating agents, such as EDTA and DPTA, when added to a system containing an appropriate bactericide, can increase activity manyfold. This effect occurs with specific combinations of bactericide and the chelating agent in specific metalworking lubricants. The chemical interactions are complex, and yet for some applications, investigation of the appropriate combination can pay off in both performance and economics.

Contamination of the metalworking lubricant by greases and/or machine hydraulic fluids may, in turn, influence the effectiveness of the antimicrobial agents. Depending upon the specific chemistry, the antimicrobial agent may be nullified as far as its effectiveness is concerned or potentiated, or have little or no effect.

BIBLIOGRAPHY

Becher, P., 1965. *Emulsions – Theory and Practice,* 2d ed., Reinhold, New York

Becher, P., ed., 1979. *Encyclopedia of Emulsion Technology,* Marcel Dekker, New York.

Bennett, E. O., 1974. Water quality and coolant life, *Lubrication Engineering,* 30(11), 549-555.

Development of Method of Evaluating Lubricants for Cold Working Metals, July 1959. Final report, WAL-TR364/35, MIT Press, Cambridge, Massachusetts.

Devine, M. J., E. R. Lamson, J. P. Cerini, and R. J. McCartney, 1965. Solids and solid lubrication, *Lubrication Engineering,* 21(1), 18-26.

Fisher, F. S., A. D. Cron, and R. G. Huber, 1982. Graphite and molybdenum disulfide – synergisms, NLGI Spokesman, September, 190-198.

Frost, H. F., 1983. *Synthetic Lubricants and Fluids Seminar,* Summary of Synthetic Lubricants and Other Functional Fluids, ASLE 1983 Annual Meting.

Gordon, I. O., 1970. Lubricants for warm working, *Metal Forming,* May, 127-132.

Iliuc, I., 1980. *Tribology of Thin Layers,* Elsevier, New York.

Kapra, P., and J. M. Martin, 1982. Boundary lubricants, a review, *Tribology International,* February, 1979, 37-42.

Lissant, K. J., ed., 1974. *Emulsions and Emulsion Technology,* 2 vols., Marcel Dekker, New York.

The Nalco Water Handbook, Nalco Chemical Co., McGraw-Hill, New York.

Parfitt, G. D., 1973. *Dispersion of Powders in Liquids,* 2d ed., Wiley, New York.

Purlsiss, B. E., 1976. The microbiology of coolant/lubricants, *Industrial Lubrication and Tribology,* March/April, 44-49.

Ranney, M. W., 1972. Synthetic lubricants, *Chemical Process Review, No. 59, Noyes Data Corp.*

Rosen, M. J., 1978. *Surfactants and Interfacial Phenomena,* Wiley, New York.

Sherman, P., ed., 1968. *Emulsion Science,* Academic Press, London.

Shinoda, K., T. Nakagawa, B.-I. Tamamushi, and T. Isemura, 1963. *Colloidal Surfactants: Some Biological Properties,* Academic Press, New York.

Swern, D., ed., 1979, 1981. *Bailey's Industrial Oil and Fat Products,* 4th ed., 2 vols., Wiley, New York.

Tanford, C., 1973. *The Hydrophobic Effect, Formation of Micelles and Biological Membranes,* Wiley-Interscience, New York.

7
Prescreening Lubricants for Metalworking

Determination of the effectiveness of a lubricant, as applied to a specific metalworking operation, depends upon many variables inherent in a given plant operation. Plant testing is expensive; thus, laboratory test methods have been developed that may adequately screen promising lubricants prior to plant evaluation, hence reducing the qualification costs. The tests are of two general types:

1. Tests based upon measurement of lubricant properties, such as film strength, wear, and friction
2. Tests carried out on equipment that may simulate an actual plant operation; for example, deep drawing, drilling, tapping, or rolling

This chapter describes some of the test procedures in each of these categories that are commonly used for screening or qualifying metalworking lubricants prior to plant evaluation.

The difficulty of screening lubricants prior to plant application and use stems from the difficulty of reproducing in the laboratory the important plant operating parameters. Nevertheless, useful operating judgments can be made prior to plant evaluation of a particular lubricant through the judicious application of laboratory screening methods. We will first discuss some of the property screening test methods that are useful in qualifying metalworking lubricants.

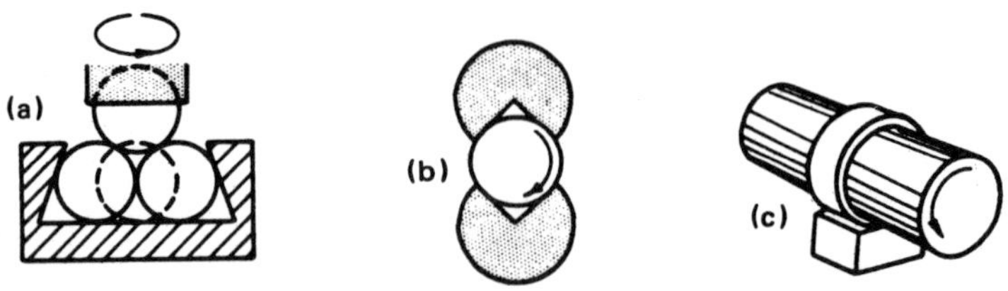

Figure 7.1 Schematic illustration of various friction and wear test apparatus: (a) four-ball tester, (b) Falex, and (c) Timken. (J. A. Schey, ed., 1970. *Metal Deformation Processes: Friction and Lubrication,* Marcel Dekker, New York, p. 287.)

7.1 LUBRICANT PROPERTY SCREENING METHODS

Many test procedures designed to measure the cohesive strength of the lubricant film, wear characteristics of lubricated surfaces, and friction as affected by lubricants have been developed. The results of such tests are then related to experience in plant operations. The film strength, wear, and friction can, with appropriate practice, be measured with good reproducibility. Translation of these data into expected operating experience is difficult and has met with limited success. Nevertheless, laboratory screening tests are useful for evaluating alternative metalworking lubricants. Some of the tests most widely used are described below.

7.1.1 Four-Ball Tests for Assessing Extreme-Pressure Properties

The measurement of friction as related to rolling, drawing, and other metalworking operations depends upon the surface of the tool and workpiece, the viscosity, and the chemical makeup of the lubricant, as well as the pressure and temperature developed during processing.

The four-ball wear test machine provides, under controlled testing conditions, a procedure for measuring friction. The machine consists of three balls touching each other, clamped together in a horizontal plane (Fig. 7.1). A fourth ball, touching the three clamped balls by positioning it between the three, is driven by a motor so that it revolves in contact withthe clamped balls. The rotation takes place within a reservoir containing a test lubricant.

A temperature-measuring device, as well as a heater, provides a method for controlling the temperature of the test fluid. The rotating ball is preloaded and then rotated for a specific time at a specific speed. At the conclusion of the test, the scar patterns developed on the balls are measured. Any stains are observed, the diameter of the scar is measured, and the coefficient of friction is calculated by dividing the tangential force by the normal force at the ball surface. The ball surface finish and chemistry, as well as lubricant concentration and temperature, may be varied along with the test conditions until results are achieved that can be correlated. Good correlation of these frictional test results have been achieved with deep-drawn cupping tests.

7.1.2 Falex Test

The Falex test consists of a method for measuring the torque (hence friction) developed during rotation of a pin between two stationary V blocks (Fig. 7.1). The pin as well as the V blocks can be made of various materials. The temperature of the lubricant bath into which the pin and the V block are placed may also be varied.

The load is applied by a ratchet arrangement. Wear measurements made above the transition pressure, that is, the load at which the lubricant film breaks down, can be a useful parameter if carried out under conditions comparable to actual metalworking operations. Particularly in those operations in which hydrodynamic lubrication can occur during at least a portion of the metalworking cycle, the mechanical strength of the film can reflect an important characteristic of the lubricant. Test procedures, testing time, testing load, increments of load, and other conditions should be varied until appropriate correlation with the desired metalworking operation is achieved. Frequently, the test procedure arrived at provides useful screening information prior to costly plant operations.

7.1.3 Timken Test

This test measures the maximum load or pressure that can be sustained by the lubricant without seizure of a steel test ring rotating against a steel test block (Fig. 7.1). When breakdown of the film occurs, a rough scar is produced on the test block. This block can be used as a comparison test piece. A series of test blocks with their score patterns may be used as a visual indication of the load-carrying characteristics of the lubricants tested.

The friction generated in the Timken test depends upon the load, temperature, sliding speed, materials in contact, and the lubricant. If the normal load and the frictional force are known, then the coefficient of friction can be determined.

7.1.4 Ring-Compression Test

The ring-compression test provides a laboratory test method that can be used to determine the frictional effects of a lubricant. The test consists of deform-

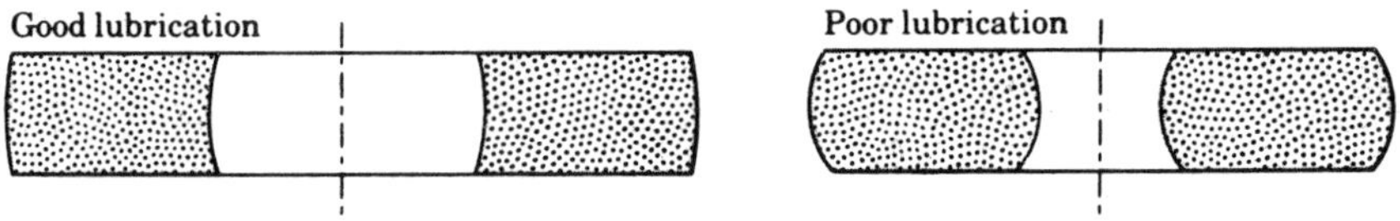

Figure 7.2 Deformation of specimens in the ring-compression test. (S. Kalpakjian, 1984. *Manufacturing Processes for Engineering Materials,* Addison-Wesley, Reading, Massachusetts, p. 203.)

ing a ring by compression between flat platens (Fig. 7.2). If there is no friction, the ring deforms as if it were a solid disk, with both inside and outside diameters increasing in proportion to their distance from the ring center. When there is measurable friction, the outer periphery of the ring is subject to greater restraint then the inner periphery. With high friction the inside diameter may actually decrease. By measuring the change in the internal diameter of the compressed ring the coefficient of friction is calculated. This can then be a relative measure of lubricant efficiency in metalworking processes. The ring (workpiece) material can be varied. The ratio of the external diameter to internal diameter and the thickness of the ring can be varied to simulate particular metalworking operations.

Thus, a technique that can relate frictional effects to a metalworking process can be very useful in screening candidate metalworking lubricants. Experience to date confirms the applicability and usefulness of the ring-compression method.

Let us now turn to a consideration of some simulation tests for qualifying metalworking lubricants.

7.2 SIMULATION TESTS

Various tests have been developed to simulate the most important metalworking operations involving metal removal and/or metal forming. Two major approaches have been taken in carrying out these tests. One involves tests carried out on laboratory equipment designed to measure some particular aspect of a metalworking operation. For example, in deep drawing of sheet metal, cup tests have been developed to evaluate the effects of various lubricants. Alternatively, an actual press operation may be used under suitable controls for lubricant evaluation. Examples of each of these approaches are reviewed in the following. These are examples selected from numerous test procedures developed over the years in an attempt to improve the cost-effectiveness of lubricant selection.

7.2.1 Metal Cutting Tests

Direct observation of lubrication during metal cutting has been carried out by the development of a transparent cutting tool. The tool, made of sapphire and placed in an appropriate holder, has been used to machine aluminum and lead, using selected lubricants. Results indicate that diffusion of the chemically active components of the lubricant during the cutting operation may not be as significant as the vapor pressure of the lubricant. Thus, surface tension and viscosity of the lubricant may prove to be more important in future developments than has been previously assumed.

Another approach is based on the use of a drill, with access ports located in the body of the drill permitting intimate contact of the lubricant with the cutting edge. This, of course, is not possible when a drilling operation is flooded externally. This special design has been used to evaluate the effectiveness of various lubricant additives. One of the interesting results of lubricant evaluations with this device has been to emphasize the importance of mist lubrication.

Direct evaluation of metal cutting fluids on production tools has a long history. More recently, machine tools have been installed in the laboratory, where control of operating conditions, tooling, and workpiece is simplified.

Drilling and tapping tests have been carried out with drilling machines or tappers selected for their rigidity. Standardized tools of specific composition have been heat treated under controlled conditions to specific structure and hardness specifications. In such testing it is important that the tool as well as the workpiece be controlled in geometry and physical and mechanical properties. The application of the cutting fluid must also be controlled with respect to its volume rate, temperature, and mode of application. With these precautions, useful evaluations have been carried out on cutting fluids for tapping and drilling.

Similar test procedures have also been developed for milling and reaming operations. Turning with single-point tools has also been used. In all these cases, feed and speed must be controlled and varied in a systematic manner so that lubrication under varying operating conditions can be evaluated.

7.2.2 Metal Forming Tests

Historically, the development of metal forming lubricants has depended upon testing in the field on actual production equipment. To improve the control of operating conditions, as well as availability of testing time during production, actual operating equipment has been set up in the laboratory for testing purposes.

Sheet-Metal Forming

In deep drawing or press operations in general, both mechanical as well as hydraulic presses have been installed in the laboratory using standard tooling. For these tests, standard tool geometries must be established, with respect to profile radius of the punch as well as the die clearance, punch diameter, and surface finish. Furthermore, the punch and die must be heat treated in a controlled manner to provide for reproducibility of both mechanical and physical properties. Application of lubricants should be controlled, utilizing either misting or airless spraying equipment, roller applicators, or any other method consistent with appropriate operating conditions. The press itself must be kept in good condition, to have repeatable bulk elastic properties and mechanical operation.

The workpiece materials must also be reproducible. The sheet metal should have reproducible mechanical properties, surface finish, and standard shape. Furthermore, the speed of testing should be controlled, the concentration of lubricant per unit area of formed part, where appropriate, must be reproducible, and the temperature must be similar for comparative testing purposes.

Alternatively, to simulate different aspects of sheet-metal forming on laboratory test devices, strip draw tests have been developed for evaluating strip-tool friction. Strip drawing through wedge-shaped dies is achieved by utilizing a modified tensile tester with the strip gripped at one end and pulled through the die. A similar arrangement may be used to pull through a die with offsets to simulate the geometry of a die with draw beads (Fig. 7.3). In both tests the resistance to sliding with alternative lubricants is measured.

Deep drawing tests may be carried out by forming cups in a standardized way, utilizing equipment that permits control of speed with a punch of fixed geometry and finish. A series of blanks of increasing diameter and coated with a lubricant are formed with a punch. The larger the blanks formed and the deeper the cup, the more effective is the lubricant. The type of blank material, size, and thickness, as well as composition and properties of the blanks, may be varied as appropriate. Die clearance and blank holder pressure, as well as size and shape of the tooling, may also be varied as desired to approximate actual production of sheet-metal parts.

In general, the limiting draw ratio (LDR) depends upon calculating the drawing ratio (DR), which is equal to the blank diameter over the punch diameter. For a given diameter of punch, the greater the blank diameter, the more difficult is the operation. The LDR is the maximum DR providing a satisfactory cup. Surface characteristics of the drawn material should be observed in terms of finish as well as the appearance of score marks.

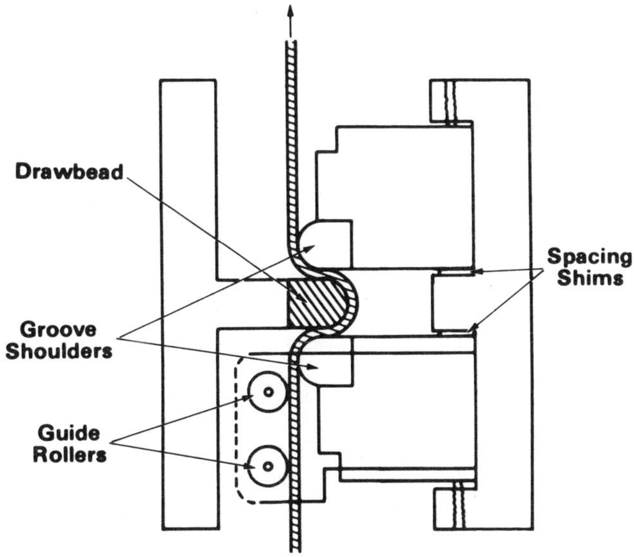

Figure 7.3 Schematic illustration of a draw-bead tester. (H. D. Nine, 1978. In *Mechanics of Sheet Metal Forming,* D. P. Koistinen and N. -M. Wang, eds., Plenum Press, New York, pp. 179–211.)

Drawing Bar and Wire

Tensile test equipment has been adapted to grip lubricated pointed wire or rod to pull it through appropriate fixed dies to measure forces and observe surfaces. Small drawbenches and/or wire drawing capstans have similarly been used where higher speeds were considered to be essential for simulation purposes. In these latter cases, as in the tensile test procedures, control of material mechanical properties as well as finish, control of die angle as well as finish of the die, and control of speed of deformation all contribute to the reproducibility of test results.

Split dies for bar drawing have been used instrumented with strain gauges to measure frictional forces as they vary with different lubricants. It is also possible to instrument the die to measure temperature. If split dies are used, then drawing can be stopped to observe the plastic flow, both prior to and through the die on the drawn material. Analysis of this flow may also give useful information about lubrication.

Rolling

Development of lubricants for rolling has been carried on with many different variations on two-high and four-high rolling mills, instrumented to measure load, temperature, and frictional forces. Control of the material to be rolled with respect to mechanical properties, as well as surface and dimensional characteristics, is indispensable to reproducible testing. The work rolls must be controlled with respect to mechanical properties and surface finish. Rolling speed as well as amount of deformation can be varied relative to strip thickness to maximize the various conditions that may be tested for lubricant suitability. With appropriate instrumentation, variations in strip thickness, roll speed, and friction or load can be measured during the rolling operation. Determination of the lubricant film thickness along with the coefficient of friction makes possible the calculation of the shear stress in the interface between the roll and strip. This, in turn, can be related to the lubrication mechanism.

Similar laboratory testing on manufacturing equipment located in the laboratory has been usefully carried out on hot extrusion, hot and cold forging, hot rolling, shape drawing, and hydrostatic extrusion.

7.3 GENERAL OBSERVATIONS

A judicious use of laboratory test methods for evaluating properties important in metalworking can frequently reduce the cost of lubricant evaluation under operating conditions. When this direct transfer of laboratory test results to operations is not effective, then laboratory control of actual production test equipment becomes necessary.

The past decade has seen accelerated development of laboratory test procedures for evaluating important lubricant criteria so that the data are directly relevant to production lubrication experience. We can anticipate continued developments in this direction. Nevertheless, because of the many operating criteria that may or may not be related directly to lubrication, that are important in designing a process, great care must be taken in transferring laboratory test data to the plant.

BIBLIOGRAPHY

ASTM D2782-74, 1974. *Measurement of Extreme Pressure Properties of Lubricating Fluids (Timken Method).*

ASTM D2783-72, 1972. *Measurement of Extreme Pressure Properties of Lubricating Fluids (Four Ball Method).*

Brown, R. D., 1969. Test methods, in *Boundary Lubrication – An Appraisal of World Literature,* Ling, F. F., E. F. Klaus and R. S. Fein, eds., American Society of Mechanical Engineers, New York, Chapter 12.

DeChiffre, L., 1980. Mechanical testing and selection of cutting fluids, *Lubrication Engineering,* 36(1), 33-38.

Doyle, E. D., J. G. Horne, and D. Tabor, 1979. Frictional characteristics between chip and rake face in continuous chip formation, *Proceedings of the Royal Society,* A366, 173.

Faville, F. A., and W. A. Faville, 1968. *Falex Procedures for Evaluating Lubricants,* Twenty-third ASLE Annual Meeting.

Friction and Wear Devices, 2d ed., 1976. American Society of Lubrication Engineers, Park Ridge, Illinois.

Male, A. T., and M. G. Cockcroft, 1964–1965. A method for the determination of the coefficient of friction of metals under conditions of bulk plastic deformation, *Journal of the Institute of Metals,* 93(38), 38-46.

Roncin, B., and M. Thebault, 1982. Simulation in the Laboratory of Industrial Deep Drawing Operations – Selection of the Lubricant, in *Lubrication in Metal Working,* Third International Colloquium, Esslingen, Germany, 25.1-25.8.

Schey, J. A., 1983. *Tribology in Metalworking: Friction, Lubrication and Wear,* American Society for Metals, Metals Park, Ohio, Chapter 5.

Siddaraj, K. V., P. N. Rao, and R. P. Arora, 1980. Some effects of mist application of cutting fluids through a pilot hole in drilling, in *Metalworking Lubrication,* Kalpakjian, S. and S. C. Jain, eds., 201-204.

8
Application of Metalworking Lubricants

Lubricant application methods can profoundly influence many technical and economic aspects of a metalworking operation. Proper choice and design of the optimum application method result in reduced overall usage of the lubricant, as well as reduced carry-off from the finished part. Frequently, the speed of the manufacturing operation can be increased through appropriate lubricant application. Tool and die life can be prolonged. Work areas may be cleaner. The workpiece will not be as slippery to handle, thus minimizing accidents. Scrap is often reduced, and workpiece surface finish is improved. Waste disposal problems and costs can be minimized. Not only is the mechanical system selected for a particular application important, but also the physical form of the lubricant (mist, flood, or coating) is important. The direction of application of the fluid to the tooling or workpiece, as well as positioning the fluid applied to tooling or workpiece, also can be critical. Figure 8.1 illustrates good and poor practices in metal cutting operations.

The lubricant delivery can influence the cooling rate, the rate of debris removed, and the location of preferred lubrication of tool or workpiece. Furthermore, the volume of lubricant delivered, as well as the rate of fluid application, can affect the metalworking process.

The importance of delivering the correct physical form of lubricant to the appropriate location at the right time, with the optimum volume and velocity at the proper temperature, cannot be overemphasized.

Some of the major lubricant application systems as they are used in important metalworking processes are described below.

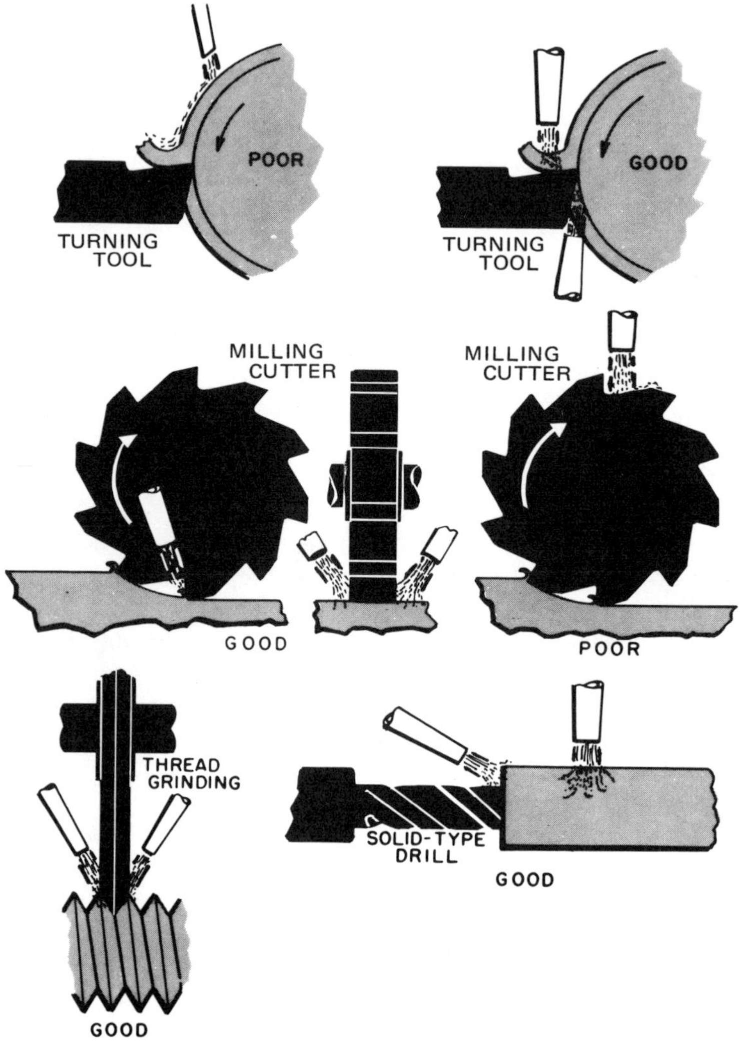

Figure 8.1 Methods showing proper and improper application of cutting fluids. (H. F. Weindel, 1981. *Tooling and Production,* November.)

8.1 ROLLER COATING

Roller coating of lubricants is widely used for lubrication of sheet metal, strip, or blanks prior to forming in a press. Roller coaters may be either powered or free turning, with and without a recirculating system, and with the lubricant applied to the roller either by appropriate submersion of the roller in the reservoir or by drip application to the roller. The roller coater can be constructed in various widths, thus providing lubricant to a very wide range of strip or blank widths. The surface of the coater can be textured or grooved to be more effective in applying the lubricant to a particular surface. Furthermore, the roller may be made from any one of a number of materials varying in hardness and surface finish, such as steel, felt, and polymers of various kinds.

The thickness of the lubricant film can be controlled and varied, depending upon the tension applied in pulling the strip or blank through the roller coater, as well as the pressure applied between the rolls. The viscosity of the lubricant also influences the film thickness. Furthermore, either top or bottom surfaces may be left uncoated as desired for a particular forming operation.

Generally, unpowered rollers are used for relatively short production runs in which the operator controls the application of the lubricant and film thickness is not critical. Powered rollers are particularly advantageous for long runs requiring maximum control of the lubricant thickness and uniformity of application.

8.2 DIP COATING

Parts for subsequent operations or slugs and billets to be cold or hot extruded may be dip coated by either hand dipping or by conveyorized movement through a bath of the lubricant. In the case of solid coatings (such as MoS_2 or graphite), coating is done by barrel tumbling. The system is simple and inexpensive to install. However, coating thickness and uniformity may be difficult to control. Furthermore, drag-out or excessive coating and waste of the lubricant may result when coatings are applied by dip coating.

8.3 AIRLESS SPRAY

For such operations as perforating sheet or plate, fine blanking, and progressive drawing or rolling, spray application of lubricants can be very effective in directing the lubricant to specific areas of the tooling and/or work material. Unlike air spray systems, very viscous lubricants may be sprayed by this method. Airless spray equipment is particularly desirable when the lubricant to be used has a tendency to mist when sprayed with an air system or, because of its chemical and physical properties, has a tendency to volatilize no matter what the application method.

By suitable choice of spray nozzle and pressure, not only may the lubricant spray pattern be controlled, but it may, by suitable timing controls, be activated only when the cycle calls for a lubricant. Most spray units have adjustable nozzles so that the volume of lubricant applied can also be adjusted.

8.4 AIR SPRAY

A lubricant spray installation powered by air may be successfully used when the lubricant does not readily mist and when its viscosity is below 300 Saybolt Seconds Universal (SSU) at 100°F (38°C). The air spray equipment is generally less expensive to install than the airless spray equipment. Nevertheless, in both cases, appropriate location of lines and nozzle is necessary to reduce spray system downtime and improve consistency of application.

8.5 DRIP APPLICATION

Drip application of a lubricant is an effective method in roll forming as well as in deep drawing. This method is useful in applying lower viscosity lubricants. However, the system design of the metalworking operation determines whether installation of the drip lubricator is practical. Most drip systems have a regulating valve, which is used to adjust the lubricant flow rate. However, this application method is not automatically initiated or stopped. Thus, generally, lubricant usage is unnecessarily high. The work area tends to become excessively covered with lubricant because of runoff. Nevertheless, the system is inexpensive and can be used effectively if short runs are anticipated and when the stock to be lubricated is small in dimension.

8.6 FLOOD APPLICATION

Flooding of the tool-workpiece area with the lubricant is widely used in many machining operations, such as grinding and drilling, as well as in some transfer press operations. The press must be set up for such application, the lubrication system of the press should be shielded, sensing systems should be waterproofed if the lubricant is water based, and a reservoir recirculating system must be available and built into the press design. However, when it is possible to use flooding, cooling as well as the availability of lubricant to points of critical need may result in improved operations. Higher press speeds can be used, and since water-soluble emulsions are generally used, lubricant film thickness is not excessive. Recirculation provides a method for maintaining proper concentration, cleanliness, and control. In metal cutting operations, frequently designed to use flooding application of lubricant, flooding also provides cooling and good access of the lubricant to the critical areas between the tool and the workpiece. In multistage operations flooding is a reasonable alternative that should be investigated.

8.7 ELECTROSTATIC AND ELECTROCHEMICAL DEPOSITION

Development of electrostatic deposition of lubricant films, from both water solutions and emulsions as well as electrochemical deposition from salt baths, have opened interesting possibilities of control of lubricant application to appropriate regions of the workpiece. Specially formulated lubricants are required.

The electrostatic method of lubricant deposition is essentially a system much like electroplating, using the lubricant as the plating bath. Both wire drawing and press forming operations can be carried out using this technique. Electrodeposition from molten salt baths has been used successfully in lubrication for warm forging of titanium alloys using a lithium chloride-potassium chloride eutectic salt mixture. Forging operations have been carried out successfully using electrodeposited coatings on slugs.

8.8 SWABBING

An effective method for lubricating the tooling in hot forging is the use of a brush or cloth saturated with the lubricant and manually rubbed over the die surface prior to forging. This method is particularly useful when suspensions of solids and liquids are used as lubricants and when the production requirement is for relatively few parts.

8.9 HIGH-PRESSURE JET

In grinding and drilling, investigations have been carried out that show the beneficial effects of using lubricants under relatively high pressures (50 to 1500 psi; 0.35 to 10 MPa). In grinding, the jet should be in a direction parallel to the movement of the wheel, thus taking advantage of the wheel rotation to ensure cleaning of the wheel, as well as cooling and lubrication of the workpiece. Previous analysis suggested that an air blanket prevented access of the lubricant to the grinding wheel. More recently, the lubricant access problem has been related to perturbation of the lubricant flow and not to one of an air film barrier. Indeed, it has been shown that very low pressures penetrate to the interface of the grinding wheel and suitably lubricate, cool, and clean the wheel if the direction of fluid flow and pressure are matched to the grinding wheel speed.

In drilling, on the other hand, drills have been designed with internal passages, with outlets adjacent to the cutting edge of the drill. Suitable lubricants fed through such a system improve lubrication, cleaning, and cooling, increasing drill life and improving hole shape accuracy.

8.10 HYDROSTATIC LUBRICATION

In wire drawing, the use of the Christopherson tube to provide thick film or hydrodynamic lubrication has been used successfully, particularly for fine wire drawing. This is a tube, placed at the entrance to the die, with internal diameter slightly larger than the entering wire diameter. The small clearance, together with high drawing speed and the viscosity of the lubricant, generates a hydrodynamic lubricant film at the die entry.

In deep drawing of sheet metal, by placing lubricant outlets in the base of the die plate at those points where lubrication is to be maximized, hydrostatic lubrication can be developed by the use of pressures at the desired locations in excess of the forming pressures required. Improved deep-drawn parts have been produced using this technique.

8.11 MISTING

A method of lubricant application that has been particularly useful in machining (drilling and cutoff) or perforating involves the application of a fine mist through a spray orifice to the tool. Mist generators usually have filters for filtering the air and lubricant to be misted, pressure regulator, mixing valve, discharge nozzle, and a fluid reservoir. Air pressure between 10 and 80 psi (70 to 550 kPa) is required, depending upon the desired particle size of the misted lubricant. Use of a cutting fluid is minimized while an appreciable amount of air is used.

Water-based fluids are particularly effective when misted. The large surface-volume ratio of each liquid particle maximizes the possibility of rapid fluid vaporization. Formation of this vapor phase not only increases the cooling effects of the lubricant, particularly cooling related to evaporation of the water, but also promotes penetration of active molecules at the tool-chip and tool-workpiece interfaces.

8.12 PRECOATING

In high-volume sheet-metal forming operations a lubricant coating is applied at the steel mills, which after some years of development shows promise. This method eliminates, for large production runs of formed sheet-metal parts, the need for lubrication at the press. The coatings are uniform in thickness, provide rust protection characteristics, have good lubricating properties, and yet they may be removed in an alkaline cleaning system without undue cost. When part production is high and shape is uniform, precoating can be a cost-effective means of lubricant application.

8.13 MANUAL METHODS

With all the technology that has been developed for application of lubricants, wiping with a rag soaked in a lubricant or brushing with a brush saturated in the lubricant remains a widely used method, minimizing capital costs for equipment. Although this method is difficult to control, it is cost effective for short-run production and when control of lubricant thickness and location is not particularly critical.

8.14 STUFFING BOX

In wire drawing the use of stuffing boxes, through which the wire passes submerged in dry soap prior to entering the die, provides cost-effective lubrication. The development of improved liquid lubricants requires flooding procedures rather than the use of the stuffing box. However, soap lubrication is widely used and is the standard method of application in the wire industry.

In hot extrusion, modified stuffing boxes are used for application of low-melting glass cake to heated billets, which are then rolled through glass powders. The method is effective, although lubricant thickness is difficult to control.

BIBLIOGRAPHY

Ivaska, J., 1972. Slip shod lubricant application can be costing you money, *Metal Stamping,* September, 7–10.

Owais, M. K. A., and B. Fogg, 1980. Recent developments in the analysis and applications of hydrostatic lubrication in sheet metal forming, *Institution of Mechanical Engineers,* C140/80, 1-8.

Rowe, S. A., 1980. Selective lubrication by electrochemical deposition for sheet and bulk forming operations, *Institution of Mechanical Engineers,* C141/80, 9–13.

Schey, J. A., 1983. *Tribology in Metalworking: Friction, Lubrication and Wear,* American Society for Metals, Metals Park, Ohio.

9

Effect of Process Fluids and Maintenance Lubricants on Metalworking Lubrication

Process fluids, such as mill-applied rust preventatives, machine maintenance lubricants, and hydraulic fluids, if allowed to contaminate the metalworking lubricant, may affect the performance and life of metalworking lubricants. These effects are discussed here to develop an understanding of the many complex interrelationships that may influence the performance of the lubricant.

9.1 PROPERTIES

It is first necessary to appreciate some of the properties of the materials that may be present as contaminants of the metalworking lubricant. These properties are described below.

Gear Oils

Napthenic or paraffinic-based-oils may have sulfur-containing compounds, chlorine-containing compounds, and heavy metals, such as lead, as well as fat or more purified esters present as components. Viscosity can vary depending upon the equipment and the gear to be lubricated, ranging from approximately 100 SUS (20 cSt) at 100°F, to 10,000 SUS (2200 cSt). Gear oils are generally very dark in color. During oiling of the bearing, new gear oil or, alternatively, old gear oil extruded from the bearing case may build up on the ways of the metalworking equipment, the blanks being formed, or the bar-stock being machined, and/or it may fall into the metalworking lubricant itself.

Way Lubricants

Naphthenic and paraffinic oils may serve as the base-oil. Sulfur and chlorinated organic molecules may be present as additives, and furthermore, high-molecular-weight polymers are often used to ensure good adhesion to the metal. Excessive lubrication of the way, or splashing of the metalworking fluid on the way, results in contamination of the metalworking lubricant. Furthermore, becuase way lubricants are formulated to adhere to metallic surfaces, they tend to be sticky and are prone to buildup on machine surfaces

Hydraulic Fluids

Oil-based hydraulic fluids may or may not contain rust-preventative additives, antifoaming additives, viscosity index (VI) improvers, and extreme-pressure additives. Invert-emulsion hydraulic fluids, with water serving as the discontinuous phase in an oil-continuous phase, may contaminate the metalworking fluid with surfactants, water, and the other operating additives with rust protection and EP characteristics. Synthetic hydraulic fluids, primarily of the polyglycol type dissolved in water, will contaminate the lubricant with the polyglycol as well as water and other additives. Frequently, leaks in hydraulic systems are not readily observable; thus, volume changes as well as chemical changes can occur in the lubricant.

Rust Preventatives

Rust preventatives applied to the raw materials prior to forming or machining may contain solvents, generally petroleum solvents, water, mineral oil, or synthetic fluids, such as a polyalphaolefin. Active ingredients may be both organic or inorganic surface-active materials. They may be present as components of a slushing oil, or alternatively, they may be applied to parts that have been machined or formed that are still wet with the metalworking fluid, prior to their storage or shipment for further processing.

Grease

Many of the greases used for lubrication of metalworking equipment are water insoluble and may contain inert components, such as sodium carbonate. Structurally, greases are composed of a soap infiltrated with oil or a synthetic fluid. Soaps may be very complex in structure. Aluminum, barium, calcium, or lithium soaps are most commonly used. Furthermore, the grease may contain inert materials, such as bentone, as well as heavy metals, such as molybdenum, lead, and/or copper, either as metal particles or as metallic sulfides.

Pickle Liquor Residue

Most of the metals suitable for forming or machining are generally processed through a pickling or shot-blasting cycle to remove rust and scale. Salts of the acids used for pickling, as well as impinged shot along with oxide dust, contaminate the workpiece surface. Lime or phosphate coatings are used to protect steel surfaces from rusting prior to parts making. These coatings provide a film that prevents metal-to-metal adhesion, as well as a base for lubricant adhesion. Phosphate coatings are generally acidic. The residue may tend to help destabilize an emulsion. Lime, on the other hand, is basic and is often used to neutralize residual acids adhering to the metal surface after pickling. This basic material can increase the propensity of the lubricant to initiate skin irritation.

Water

Contamination by water has many implications because of water hardness and specific anions ($SO_4^=$ and Cl^-), as well as cations (Al^{+++}, Ca^{++}, Ba^{++}, and Fe^{++} and others) that may be present, the pH, and bacterial content of the fluid. Hard-water soaps may form with some metalworking fluids and deposit on tools and parts. Emulsions may be destabilized and split. Bacterial infestation may be accelerated.

9.2 PROCESS IMPLICATIONS

Dermatitis

Particularly in the case of cutting fluids, dermatitis may result if chlorinated and/or sulfurized products contaminate the fluid, if pH due to the presence of pickle liquor residues is reduced, and/or reactions between additive systems present in the contaminants and the metalworking fluid occur. Microbiological organisms present in water or introduced from other sources may degrade the lubricant, reduce the pH, and thus encourage an outbreak of dermatitis.

Lubrication Effects

Any one of the possible contaminants just described can influence the chemistry of the metalworking lubricant. New compounds may be formed, and original compounds may lose distinctive properties. The specific lubrication characteristics may be altered. The wetting characteristics of the lubricant may change as part of this process. The presence of grease, if it enters the application side of the metalworking lubricant cycle, may form an undesirable deposit and/or fundamentally change lubricating characteristics.

Rust and Staining

If the pH is changed, rust protection may be decreased. Furthermore, the presence of additives in the gear and way lubricants, or hydraulic fluids and rust preventatives, can lead to deposits that stain and may, in turn, be the precursor of rust.

Microbiology

The unwanted presence of water in an oil-based or a solid lubricant can introduce biological growth where none previously existed. Furthermore, any of the contaminants can influence both the life and the activity of biocides that may have previously been used to minimize biological effects in water-based lubricants.

Deposits

Reactions between lubricants and contaminants may often lead to by-products that deposit on the surface of finished parts as well as on the metalworking equipment. The deposit buildup on parts may seriously interfere with subsequent cleaning and/or finishing operations, such as welding or painting. Deposits on tooling can interfere with part production and its quality.

Cooling

Contaminants may also change the heat-transfer characteristics of metalworking lubricants. For example, contamination of emulsion cutting fluids by hydraulic fluids or way lubricants and gear oils can significantly reduce the effective cooling desired in the cutting fluid. This is due to the low thermal conductivity and specific heat of oils as compared with that of water-based fluids.

Emulsification

In oil-based lubricants, introduction of surfactants into the fluid can, with subsequent water contamination, seriously degrade the properties of the fluid. Furthermore, in emulsions the introduction of additives and/or surfactants may change emulsion particle size and operating characteristics. In the case of solutions, introduction of such foreign chemicals can either reduce or increase the solubility of oils in water-based solutions, as well as change the suspension characteristics of the fluid.

Stability

In metalworking emulsions the chemical contaminants discussed above can, in extreme cases, break the emulsion. More frequently, the emulsion characteristics can again be changed. Particle size and wetting of metal surfaces may be altered.

Foaming

Particularly in the case of solutions and frequently in the case of emulsions, contaminants that are introduced can affect foaming characteristics, ranging from extreme foam formation, where no foam existed before, to reducing the foam that is initially present. These effects are due to chemical as well as physical interactions.

Metal Fine Suspension

In recirculating systems, metal fines that are produced in the course of processing should be removed prior to reuse of the lubricant. The separation of metal fines from the lubricant can often be accelerated through appropriate chemistry, that is, through reducing the wetting of the fines by the lubricant. The chemistry of lubricants can be changed by freshly generated fines. The active surfaces serve to catalyze chemical reactions in the fluid.

Color

Frequently, introduction of contaminants changes normal appearance and color. This change, noted by operators if no one else, can lead to real as well as imagined difficulties in operation, such as skin irritation, objectionable odor, and poor tool life.

Acidity

If the pH of a water-based lubricant is reduced, the potential for corrosion is increased. On the other hand, if the pH is raised, the potential for dermatitis can increase. Thus, changes in pH not only affect performance but may reflect the presence of various contaminants. Systematic evaluation of pH can be useful in controlling the performance of the lubricant.

Viscosity

As important in metalworking operations as any other characteristic of the lubricant is its viscosity, particularly the viscosity at the temperature and pressure in effect during metalworking. The viscosity of oils can be changed by the presence of water as well as other contaminants, such as grease, and way and gear oils. The viscosity of water-based fluids can similarly be affected by changing water concentration, as well as through the contamination of other fluids with higher or lower inherent viscosity. In addition, chemical interactions may also affect viscosity; this effect can be to either increase or decrease the viscosity of the fluid under the operating conditions.

Safety

If deposits build up on equipment and/or the floor and on formed parts, handling can become more difficult, operators may slip, and part transfer can be dangerous. Furthermore, if contamination leads to fumes, smoke, or airborne irritants of any kind, operator errors can be compounded. If these occur over and over, then operators of equipment may experience undesirable health effects, such as skin irritation and lung damage.

Recycling

The economics of metalworking lubrication may justify recycling of lubricants. Recycling often entails the use of settling tanks where sludge is accumulated and fines and chips accumulate at the bottom. It may often involve skimming of surface oils and tramp oils. In both these cases, introduction of extraneous contaminants can not only affect the volume of these materials, which must subsequently be disposed of, but increase the time as well as cost of separation.

Disposal

If the metalworking lubricant must be disposed of, its chemical composition may influence the mode of disposal and the cost involved. If the contaminant is water, it is more economical to separate the water prior to disposal and thus ensure its purity. If the water-based lubricant is, on the other hand, contaminated by oil or oily layers, these oily layers must be separated from the water-based fluid and disposed of separately. In all cases, contamination of the lubricant in the plant can be expected to increase the cost of disposal.

In conclusion, the wide array of chemical and physical interactions that is possible in a metalworking lubricant contaminated by a maintenance lubricant or process fluid should now be apparent. There is no aspect of lubrication that is not potentially affected by contamination. Frequently, the effects are not of great significance and do not interfere with the metalworking operation. For example, when the effective EP is increased or the viscosity is increased, metalworking operations may improve for a period of time, until the optimum conditions have been exceeded; then, operating efficiencies abruptly drop off. In all cases, however, it is the lack of control inherent in the presence of contaminants that most seriously interferes with efficient metalworking operations. In many cases contamination can abort the operation or seriously increase costs.

BIBLIOGRAPHY

Boner, C. J., 1954. *Manufacture and Applications of Lubricating Greases,* Reinhold, New York.

Metals Handbook, 9th ed., vol. 5, *Surface Cleaning, Finishing and Coating,* 1982. American Society for Metals, Metals Park, Ohio.

Schey, J. A., 1983. *Tribology in Metalworking: Friction, Lubrication and Wear,* American Society for Metals, Metals Park, Ohio.

Wills, C. G., 1980. *Lubrication Fundamentals,* Marcel Dekker, New York.

10

Control of Lubricants: Test Methods, Plant Tests, and Laboratory Control

Metalworking lubricant properties may be evaluated and controlled by the use of suitable physical and chemical testing procedures. These tests should be carried out in a systematic way, on incoming lubricants as well as on lubricants during use. Systematic testing records can serve as a quality check on the supplier as well as promote maximum advantage in use. Production test data obtained on a regular schedule can serve to control not only the lubricant, but also the process. Changes in process parameters can often be detected in the course of systematic lubricant testing.

The economics of lubricant use may be influenced singificantly by the cost of lubricant control as well as the cost of disposal. Disposal costs may often be reduced if lubricant life is extended by suitable reclamation and control procedures.

Control of lubricants is dependent upon selection of proper testing methods and procedures. This selection is influenced by the chemistry and properties of the fluid, as well as by the type of metalworking process. For example, viscosity may be a critical characteristic of the lubricant used for deep drawing sheet steel. In central coolant reservoirs the pH as well as concentration of the fluid must be controlled. In hot forging of heat-resistant alloy steels, the solids content of the lubricant must be maintained and controlled.

The control of metalworking lubricants can be discussed in terms of control of lubricant quality prior to use as well as during use. Table 10.1 summarizes some typical test methods that are used to control major lubricant classes: (1) oils, (2) solvents, (3) emulsions, (4) solutions, (5) solids in suspension, (6) pastes, and (7) solid coatings.

Table 10.1 Testing Methods for Metalworking Lubricants

	Lubricant system						
Test Methods	Oil-based	Solvent-based	Solids	Water-based emulsions	Solutions (water)	Coatings	Pastes
Particulate concentration	X			X	X		X
Emulsifier level				X			
Viscosity	X			X	X		
Sulfur	X	X	X	X	X		
Chlorine	X	X	X	X	X		
Particle size				X	X		
Solids content	X	X		X	X	X	X
Color	X	X		X	X		
Saponification number	X	X		X	X		X
Evaporation	X	X					
Flash point	X	X					
Molybdenum disulfide		X	X				
Graphite			X				
Carbonate			X				
Stearate			X				
Water content	X	X		X	X		X
Hard-water tolerance				X	X		
Stability	X	X		X	X	X	X
Bacteria, mold, fungus				X	X		X
Amine				X	X		
Phosphate				X	X		
Corrosion				X	X		X
pH				X	X		
Conductivity				X	X		X
Concentration				X	X		
Odor	X	X		X	X		
Coating thickness	X	X		X		X	
Foaming				X	X		

This chapter is devoted to a description of test procedures useful for controlling lubricant quality and performance. The tests are divided into two groups: ASTM (American Society for Testing and Materials) standard test methods, best used by trained personnel in a laboratory, and test procedures that may be used by plant personnel during plant operations.

10.1 VISCOSITY

Viscosity is probably the single most significant physical property of a metalworking lubricant. It is a measure of the flow characteristics of the lubricant. The thicker the lubricant, generally the higher is its viscosity and the greater its resistance to flow. The mechanics of establishing a proper hydrodynamic lubricating film depends largely upon viscosity.

To evaluate the kinematic viscosity (a measure of the resistive flow of a fluid under gravity) of an oil-based lubricant, any of several test standards may be applied. Although these tests differ in detail, they are essentially the same in principle. They all measure, with a viscometer, the time required for a specified quantity of oil at a specific temperature to flow by gravity through an orifice or constriction of specified dimensions. The thicker the oil, the longer the time for passage.

Close control of oil temperature is important. The viscosity of any petroleum oil increases when the oil is cooled and decreases when it is heated. Therefore, the viscosity of the oil must always be specified with respect to the temperature at which it is determined.

ASTM D445–446 pertains to test methods that describe the instrumentation and technique for measuring kinematic viscosity. Viscometers are glass tubes of various designs (such as Cannon-Fenske and Ubbelohde) that permit flow under controlled conditions. This flow is translated into a viscosity number at a specified temperature. Kinematic viscosity is expressed in centistokes (cSt).

ASTM D88 describes another method that is commonly used for measuring viscosity using a Saybolt Universal viscometer. Saybolt Seconds Universal (SSU) are the units of measurement used to report viscosity. This method is based on flow of the lubricant into a calibrated receiver under controlled temperature. The time is measured for a given measure of viscosity. Saybolt viscosity measurements are not as precise as kinematic viscosity measurements, but they are widely used in the United States and are adequate for most metalworking fluids.

Since viscosity of metalworking lubricants is frequently the most significant single characteristic of the fluid, speed, load, and temperature conditions during metalworking operations should be related to the viscosity of the fluid to select higher viscosities when they are required to impart the necessary film strength. Low speeds, low pressures, and high temperatures require oils of higher viscosity. An oil that is too light may lack the film strength necessary to carry the load and give adequate protection to the tooling. Tool and die design and part configuration may also dictate selection of viscosity to provide the necessary mechanical characteristics desired in the lubricant.

Indirectly, viscosity is related to the volatility of the fluid: the lower the viscosity, the higher the volatility, and the more susceptible the fluid is to evaporation under high temperature operating conditions. Thus, although perhaps desirable when minimum residual film is desired on the finished part (as in painting, soldering, and welding), volatility may be detrimental to air quality and may cause objectionable odor and degradation of lubricant performance.

10.2 COLOR

The color of an emulsion can be a useful guide to the condition of the emulsion. As the emulsion ages, the particles, namely, the oil phase, generally increase in size. This increase in size leads to a deepening and darkening of the color. However, no two people have quite the same perception of color. Hue, intensity, tone, purity, wavelength, opacity, and brightness are all directly or indirectly associated with color. On the other hand, color evaluation can often be an effective method for quantifying incoming lubricant quality, predicting the useful life of the emulsion, as well as its degradation or contamination during use. Operating experience can often be translated into color standards reflecting chemical or physical changes in the lubricant.

Most color tests involve transmitted light through standard reference materials at a given depth or thickness of the sample. The two light fields are compared visually, and adjustments are made until a match is obtained. Two of the more common color tests are the Saybolt test (ASTM D156), in which the volume of the sample is varied until the fields of the known and unknown fluids match, and the color test (ASTM D1500), in which light is transmitted through a given depth or thickness of the sample and matched by using a series of colored glasses. When the operator obtains a match, a color value is recorded.

10.3 SULFUR

Oxides of sulfur are produced by burning a sample of predetermined weight in a combustion bomb containing oxygen under pressure (ASTM D129). After burning is completed and the pressure is released, the interior of the bomb is washed with distilled water and the washings are reacted with barium chloride to precipitate barium sulfate. The sulfur content is determined gravimetrically by weighing the precipitate.

Active sulfur is determined by mixing copper powder with the sample and heating it to 300°F (150°C). The resulting copper sulfide is filtered out and separated from the original sample. The remaining sample is inactive sulfur, which can be determined by the procedures described in ASTM D129. The active sulfur is considered to be that proportion of the sulfur that reacts with the metal surface of both the blank from which the part is to be formed and the tooling. Tungsten carbide tooling may be degraded by chemical attack of the cobalt binder by sulfur. On the other hand, sulfur complexes formed on the surface of the formed metal are effective boundary lubricants.

10.4 CHLORINE

Chlorine-containing compounds also form complex chemical species that react with the metal surface being formed, resulting in a boundary layer that is effective in improving tool life and surface integrity of the part.

In this case, a procedure similar to ASTM D129 is used, except that the chlorine compounds, which result from oxidation in a combustion bomb, are reacted with sodium carbonate. The percentage of chlorine is determined from the precipitated silver chloride (ASTM D808).

Both sulfur- and chlorine-containing compounds are common additives to metalworking fluids. Unless controlled, they can cause rusting, degradation of the tooling, and dermatitis.

10.5 FAT

Fatty compounds are common additives in metalworking fluids since they serve to wet metals at high temperatures and provide good adhesion to the metal surface. A weighted sample of the fluid is dissolved in metal ethyl ketone, as described in ASTM D94. This solution is mixed with a measured amount of standard potassium hydroxide solution. The unconsumed or reacted KOH is determined, the consumed quantity of KOH determines the saponification number. A given saponification number serves as the basis for a calculation of the percentage of fat, if the fatty additive is known. Caution must be exercised in carrying out this test because sulfur, phosphorus, halogens, and certain other elements can consume alkali and acids and distort the test results. ASTM D128 describes a gravimetric method for determining the fat content of an unknown composition when the saponification number is low and the fat content is questionable.

10.6 CORROSION

Copper and copper alloys are often components of equipment, or they are used in making parts. An appropriate metalworking lubricant must not stain or corrode these components. ASTM D130 describes a method for determining the activity present in the lubricant and its effect on corrosion.

The method consists of preparing a copper strip by cleaning and polishing it so that no discoloration or blemishes are visible. The fluid to be tested is then placed in a test tube, the copper strip is immersed in the fluid, and the test tube is stoppered with a vented stopper. The tube is held in a water bath for 3 hr at a temperature of approximately 212°F (100°C). At the end of the exposure period, the strip is removed, the solution is wiped off, and the strip is matched with one of the standards. The standards consist of four classifications of degree of attack, from slight tarnish, a light orange in color, to corrosion, in which black, dark gray, or brown spots appear on the surface.

There are many variations of this test. Time of exposure, temperature of the sample, and interpretation of the appearance of the exposed sample are all subject to variation and should be standardized as much as possible. Interpretation of the results is dependent upon the end use of the copper alloy and its requirements with respect to color, finish, and lack of corrosion.

ASTM D1748 is a method that has become commonly used for evaluating the corrosion of steel as it may be affected by lubricant additives. An appropriate series of steel or other metal panels is coated with the lubricant to be evaluated. The panels are then suspended in a humidity cabinet, maintained at approximately 100% humidity and at a temperature of approximately 120°F (49°C). Test results are determined through evaluation of the degree to which corrosion has occurred after a specified testing time. Size and number of rust spots generally form the basis of comparison.

10.7 EVAPORATION RATE

In many applications for which solvent-based lubricants are used, the principal property of interest is the ability of the solvent to evaporate. Subsequently formed parts are then clean enough to be painted without the need for separate cleaning operations. The time required for evaporation is an important factor in most applications. Furthermore, the rate of evaporation is never constant for a mixture of compounds but varies throughout the evaporation process; this behavior is very important for some applications, such as roll forming or sheet-metal pressworking.

A thin-film evaporameter, described in ASTM D3539 is a device used to evaluate evaporation time. A filter paper wet with the test material is suspended by a spring within the evaporameter cabinet, which is maintained at 77°F (25°C) and is subjected to an established rate of flow of dehumidified air. As the test proceeds, the time is recorded for each 10% evaporated, up to 90%, and then the times for 95 and 100% evaporated are recorded. Data from this method at the 90% evaporated point are sometimes expressed as a relative quantity normalized about the evaporation time of normal butyl acetate.

10.8 FLASH AND FIRE POINTS

The flash and fire points of a solvent-based lubricant are basically measurements of flammability. The flash point is the minimum temperature at which sufficient liquid is evaporated to create a mixture of fuel and air that will burn if ignited. The fire point, however, is somewhat higher. Fire point is the minimum temperature at which vapor is generated at a rate sufficient to sustain combustion. Safety regulations and insurance rates are two operating parameters directly influenced by flash and fire point characteristics.

In the ASTM D92 test method, an oil sample is contained in a precisely specified brass cup that rests on a metal plate. Under the plate, a source of heat is provided, and a thermometer in the cup indicates the temperature of the sample. From a temperature of 100°F (38°C) or higher, the lowest anticipated flash point, the sample is heated at a prescribed rate. At each multiple of 5-degree Fahrenheit temperature rise, the lighted torch is passed over the top of the cup. The test for flash point is completed when the torch produces a flash from the vaporized portion of the sample.

The significance of flash and fire point values lies in the dissimilarity that exists in the volatility characteristics of different metalworking lubricants, which may be hydrocarbon solvent or oil based. Even among lubricants of comparable viscosity there are appreciable variations in volatility and, hence, in flash and fire points.

Since flash and fire points are also related to volatility, they offer a rough indication of the tendency of the lubricant to evaporate in service. However, the relationship between test results and volatility is by no means conclusive and should be treated only as indicative of possible performance characteristics.

10.9 FOAMING

The foaming characteristics of a metalworking lubricant reflect the effects of the additives as well as their ability to relase trapped air. Trapped air can significantly affect the operating characteristics of the lubricant. The generation of foam during the metalworking operation can reduce effective film strength, suspended metal fines and dirt, and slow the heat transfer, thus reducing tool life and impairing surface finish. Furthermore, the workpiece can be made much more difficult to keep clean.

In ASTM Test D892, a 190-ml sample of the lubricant is heated to 120°F (49°C) and cooled to 75°F (24°C) before pouring it into a graduated cylinder, immersed to at least the 90-ml mark in a 75°F constant temperature bath. A diffuser stone is then immersed in the sample, air flow is adjusted to 90 ml/min, the air inlet tube is connected to the stone, and air is forced through the stone for 5 min. The tube is then disconnected, and the volume of foam is recorded immediately. The sample is allowed to stand for 10 min, and the volume of foam is again recorded. The test is repeated in a 200°F (93°C) constant-temperature bath. Any foam remaining after the 200°F test is collapsed by stirring. The sample is removed from the bath and cooled to a temperature below 110°F. The cylinder is placed in the 75°F bath, a diffuser stone is inserted, and the procedure is repeated.

The test results are reported as volume of foam in milliliters at the end of the 5-min blowing period and at the time of the 10-min settling period for the three different test conditions. The volume of foam at the end of the blowing period is a measure of the foaming tendency of the lubricant; the foam volume at the end of the settling period, which is usually zero, is a measure of the stability of the foam.

Many factors affect the foaming qualities of emulsions and solutions. Both the compositions and the contaminants can directly influence this characteristic.

10.10 PARTICULATE CONCENTRATION

ASTM D273 describes a method of measuring the amount of material deposited on a filter under controlled conditions, thus permitting an evaluation of the buildup of particulate matter in a metalworking fluid. Particulates can result in aggravating skin conditions, bacteria-assisted degradation of the fluid, and breakdown of nitrogen-containing additives and can have deleterious effects on tool life and surface finish of parts.

10.11 EMULSION STABILITY

Soluble oils or emulsions that are commonly used in metalworking operations may separate into their components if not formulated for particular plant conditions (such as hard water). Water condition is of particular concern since the chemistry of dissolved anions and cations may split the emulsion. Furthermore, an emulsion suitable prior to use may become unstable during use due to a buildup of tramp oil (hydraulic fluid or gear lubricant), changes in temperature, or biologically induced degradation.

ASTM D1479 describes a method for determining emulsion stability in which an emulsion of suitable concentration is made up from the soluble concentrate and water of a standard hardness. The emulsion is stored for 24 hr, after which the bottom fifth is separated for determination of its oil content. Comparing the oil concentration stored and the freshly prepared emulsions gives the percentage of oil depletion, thus a measure of emulsion instability.

10.12 ADHESION OF COATINGS

Coatings preapplied to the workpiece have been successfully used on sheet metal prior to forming. The adhesion of these coatings is an important attribute, which should be controlled. ASTM D2510 describes a test method that consists of applying the coating to an aluminum test panel, which is subsequently immersed in the test fluid for 24 hr and is then wiped dry. A strip of masking tape is pressed onto the panel and is then removed abruptly. The ease of film or coating removal resulting in exposure of the surface of the metal strip is a criterion of failure of the coating to adhere to the metal surface.

10.13 HYDROGEN ION CONCENTRATION

Frequently, plants engaged in the use of metalworking lubricants do not have appropriate laboratory facilities. Under these circumstances simplified test procedures must be used.

Monitoring the pH of the metalworking fluid is a simple method of determining the stability of the fluid. The pH or hydrogen ion concentration of a solution or emulsion is a measure of its overall acidity or alkalinity. A change in pH indicates that the hydrogen ion equilibrium of the fluid is disturbed and that the components in the fluid may be undergoing degradation, either by biological or by chemical mechanisms. Conversely, a change in pH may indicate gross contamination resulting from external sources. The pH can be measured using a standard pH meter or by the use of pH paper. Alkalinity or acidity can also be determined through a standard titration procedure using phenolphthalein indicators. Monitoring the pH on a daily basis helps keep track of the lubricant stability under plant operating conditions.

In plant operations in which a water-based fluid is recirculated, a record of lubricant characteristics should be kept on an ongoing basis to aid in the evaluation of the conditions of the fluid. These tests, which are simple in execution, justify their use very quickly in extending the useful life of the metalworking fluid.

10.14 FREE ALKALINITY

This test is used to determine the reserve or excess alkalinity remaining in the lubricant. Most emulsions are formulated with some excess alkalinity for improved stability, rust protection, and bacteria control. The actual amount of free alkalinity present appears to be of secondary singificance. It is important to detect changes in the amount of free alkalinity present and to determine what is causing the change.

The free alkalinity test is performed by titrating a 10-ml sample of lubricant with 1/10 normal acid to a methyl orange end point. Normally, the alkalinity will gradually increase with the use of a coolant due to the accumulation of carbonates from makeup water. A rapid increase in free alkalinity is an indication that a basic material (contaminant) is entering the fluid. On the other hand, bacterial attack and breakdown of additives may reduce alkalinity. Skillful interpretation of these possibly competing effects can be very useful in maintaining a given metalworking operation under proper control.

10.15 SEPARATION TEST

The amount of unemulsified or weakly emulsified oil present in a lubricant emulsion is important. A suitable test for the stability of the emulsion can be performed by filling a graduated cylinder to a 110-ml mark with a sample of lubricant and allowing it to stand for 24 hr. The amount of oil that has separated from the emulsion and floats to the top and the amount of cuff is recorded. Cuff (a creamlike layer that forms just below the oil) consists of partially emulsified oil that floats at the top but has not completely separated from the water phase. The greater the amount of free oil and cuff, the less stable in the emulsion.

10.16 DEPLETION TEST

A test procedure used to determine the degree of instability or deterioration of an emulsion is useful in the selection process for a new metalworking emulsion. Such a test can also be used to evaluate the control measures used to maintain a given emulsion or synthetic fluid. This test can be performed in conjunction with a 24-hr standing separation test. It is a slight modification of ASTM D1479. After recording the amount of free oil and cuff, 20 ml of coolant is drawn from the bottom of the graduated cylinder with a 20-ml pipette, and the concentration is checked by breaking it with sulfuric acid. This concentration (C_B) is compared to the initial concentration (C_I) of the sample run the previous day. The percentage depletion (D) or deterioriation is calculated by the formula

$$D\ (\%) = \frac{C_I - C_B}{C_I} \times 100 \qquad (10.1)$$

The higher the percentage of depletion, the poorer is the emulsion stability of the fluid. Babcock bottles (flasks with calibrated narrow necks) can assist in making as accurate a volume reading as possible and are superior in this regard to graduated cylinders. Deterioriation of an emulsion prior to its use (forming of a cuff or oil layer) can influence the choice of a particular metalworking emulsion. Generally, during use, this deterioration is accelerated, not retarded.

10.17 CONCENTRATION

The most significant attribute of a water-based emulsion or solution is its concentration during use. The concentration of active ingredients directly influences performance characteristics. If this concentration varies in an unknown manner, control of the operation can be difficult, if not impossible. Application methods, contamination effects, and other operating parameters can influence performance significantly, but concentration ranks among the most important attributes of the metalworking fluid.

Emulsions are mixtures. When the concentrate is mixed with water prior to use, the determination of the concentration of the concentrate is relatively straightforward. Frequently, the water added to the concentrate is contaminated with salts, bacteria, and/or other chemicals to such an extent that there is immediate interference with any analysis of the resulting emulsion.

It is with use that the problem of concentration determination appears most often. During use, the separate components of the fluid (such as rust preventative, extreme-pressure additive, emulsifier, and surfactant) are depleted at different rates, through bacterial attack, reaction with elements in the water or metal, drag-out (lubricant adhering to material) on the parts, or degradation due to temperature and pressure of the metalworking operation. This increasingly severe change of composition with time and use can degrade the performance characteristics of the fluid.

If the compositional changes are not corrected through selective additions during use, the fluid may fail in one or more of its requirements. Parts may rust, bacteria may grow unchecked, tool life is reduced, and part dimensions and/or surface finish no longer meet specifications. Thus, control of concentration during use is very important.

Four methods of particular singificance in determining concentration in the plant, or relatively quickly with a very modest laboratory setup, are (1) refractometer, (2) emulsifier titration, (3) acid split, or (4) acid titration.

Refractometer

A physical method of monitoring metalworking lubricants that are emulsions or synthetics utilizes a small pocket refractometer. The refractive index of a fluid increases with increasing concentration and is read directly on this instrument. Unfortunately, matter foreign to the lubricant, which is soluble or suspended in the fluid, will be reflected in the reading. Since the scale on the refractometer is arbitrary, values for each lubricant tested must be established with a known series of dilutions to establish the correct concentration relationship to scale reading.

Because the method is simple and rapid, it is convenient and appropriate to use when the process sample can be kept relatively uncontaminated by tramp oil or other soluble as well as suspended matter. The instrument should be recalibrated regularly. Care must be employed in taking the sample for concentration measurement to ensure that the sample is truly representative and is not contaminated by surface tramp oil or deposited sludge.

Emulsifier Titration

Anionic or nonionic emulsifiers are most frequently used to promote emulsion stability of a lubricant. Anionic emulsifier concentration can be analytically titrated with simple techniques. Two types of anionic emulsifiers are most frequently used in metalworking fluids: alkanolamine salts of fatty acid and alkaline metal salts of a petroleum sulfonate. Both anionic emulsifier classes can be colorimetrically titrated by a cationic material using an appropriate indicator (anionic, cationic, or a mixture of these).

An emulsion using a fatty acid ethanolamine emulsifier may be titrated against hyamine (cationic titrant) in the normal two-phase system, using bromophenol blue as the anionic indicator. Another indicator that may be used consists of a mixture of an anionic (disulfine blue V) and a cationic (dimidian bromide) dye.

If a petroleum sulfonate is used as the emulsifier, it can be titrated with hyamine using a cationic indicator (methylene blue). Several methods are listed in Table 10.2.

In all cases in which titration of the emulsifier is an appropriate control procedure (anionic emulsifiers), the influence of water hardness must be taken into account. The hardness of water can seriously affect the apparent concentration determined by titration, and hence water of known hardness should be made up for dilution to known concentrations, then titrated to establish the water hardness effect.

Emulsion Acid Split

Determination of the water content of an emulsion can be carried out if the emulsion is destabilized and separates into layers, one of which is water. A sample of the emulsion is separated from any tramp oil or sediment, and a strong mineral acid (H_2SO_4) is added. After heating the mixture for an hour or so, the once opaque emulsion separates into a lower layer (H_2O), with an upper layer consisting of the oily material present. The volume of the oily material can then be related to the water volume. This reflects the lubricant concentration. Any oil contamination of the system from machine lubricants, for example, will distort the results of such an oil split, since the oil volume will reflect not only the oil of the metalworking emulsion, but also the tramp oil present. Nevertheless, the method is useful under plant conditions.

Acid Titration

Water-based metalworking lubricants usually have a pH between 8.5 and 9.5. Because of its alkalinity, the solution or emulsion may be titrated with a dilute mineral acid to a predetermined pH value and then compared with a standard alkaline solution (dilute NaOH) to establish an alkaline equivalent that rises with any change in the pH of the metalworking lubricant. Since the majority of components in a fluid contribute (although not in equal proportions) to the overall alkalinity of the mix, a periodic check of the alkaline equivalent of the fluid serves to monitor any change in concentration. Contaminants and biological attack, as well as chemical degradation, may also influence the alkaline equivalent. Thus, the interpretation of the results of such

Table 10.2 Chemical Methods to Measure Cutting Fluid Mix Concentration

Titrant	Indicator	Solvent	Buffer	Shaking method	End point	End-point color due to	Remarks
Hyamine 2389,[a] Hyamine 1622[a]	Bromophenol blue (Anionic; BPB)	Carbon tetra-chloride	Boric acid, potassium chloride, sodium hydroxide, pH 9 ± 0.1	10 slow inversions	First trace of blue color in solvent	Solvent soluble reaction product from excess hyamine and bromophenol blue	Sensitive to hardness of the mix water
Hyamine 1622	Mixed indicator; disulfine blue V + dimidium bromide	Chloroform + isopropanol (95:5)	Boric acid, potassium chloride, sodium hydroxide, pH 9 ± 0.1	Hard shaking	Pink solvent phase turns blue/green	Solvent soluble reaction product from excess hyamine and disulfine blue V	Not sensitive to hardness of the mix water (CIA Standard)

HTAC[b]	Bromocresol green (anionic)	Chloroform + isopropanol (2:1)	Sodium chloride, sodium phosphate, sodium hydrogen phosphate, pH 9	Hard shaking	Water and solvent phase of equal blue intensity	Solvent soluble reaction product from excess HTAC and bromocresol green	Not sensitive to hardness of the mix water
HTAC	Methylene blue (cationic)	Chloroform	Sodium sulfate, sulfuric acid, pH 2	Hard shaking	Water and solvent phase of equal blue intensity	Liberation of water-soluble methylene blue from the solvent-soluble emulsifier-methylene blue complex by HTAC	Suitable for sulfonate emulsifiers, hard water insensitive

[a]Hyamine 2389 = methyldodecylbenzyltrimethylammonium chloride + methyldodecylxylylenabistrimethylammonium chloride.
Hyamine 1622 = *p-tert*-octylphenoxyethyldimethylbenzylammonium chloride.
[b]HTAC = Hexadecyltrimethylammonium chloride.
Source: G. J. P. Beckert, 1982. In *Lubrication in Metal Working,* vol. 2, Third International Colloquium, Esslingen, Germany, p. 105.8.

tests are best made using specific operating experience to influence interpretation of the data.

10.18 CORROSION TESTS

Because one of the attributes of water-based fluids particularly of importance to the selection process, as well as during operation, is the response of equipment, tooling, and workpiece to the fluid with respect to corrosion, testing may be carried out simply through the use of cast iron chips. The chips are solvent cleaned then subsequently coated with the solution or emulsion in question, drained, and observed after 24 hr, 48 hr, and so on, to determine the extent of rusting.

A similar test may be carried out on ground and polished coupons. In this case, the surface is again cleaned and covered with a sample of the emulsion or solution and is set aside to be observed for the onset of rusting.

In both these cases (that is, the cast iron chip as well as coupon tests) the extreme sensitivity of cast iron to corrosion provides acceleration of the test. This accelerated corrosion may serve as a check on the possible deterioration of corrosion protection during use.

If a humidity cabinet is available or one can be constructed that will provide on the order of 100% humidity at some controlled temperature, the cleaned test panels of the material with the corrosion protection in question may be suspended in such an atmosphere and observed for the onset of corrosion.

The relationship between any of these tests and actual operating experience must be developed through testing in the plant under specific operating conditions of temperature, environment, fluid contamination, and workpiece material changes, as well as machine and tool condition. Nevertheless, experience can lead to the development of an evaluation procedure for new fluids as well as in-process changes that may relate to corrosion.

10.19 PARTICLE SUSPENSION

For lubricants that consist of graphite and/or molybdenum disulfide suspended in a fluid medium, the stability of the suspension, or lack of it, is important. A sample placed in a glass cylinder for observation of suspension stability over time can provide the basis for evaluation of a given fluid suspension. If a centrifuge is available, centrifugation of a sample of the suspension can serve to accelerate precipitation of the particulates in the suspension. By recording the time, temperature, and centrifugal force exerted during the test, a comparative rating of stability for alternative fluids can be obtained. This rating can be used for in-process control as well as for qualification of new fluids.

10.20 CLEANING AND WELDING

The potential of a deleterious effect of a chosen lubricant on subsequent cleaning or welding operations should be evaluated prior to its acceptance and then, periodically, during use of the lubricant. Simple tests using the plant cleaning and welding procedures on lubricated stock can provide a basis for evaluation of the ease or difficulty of each of these operations as they may be influenced by the lubricant.

10.21 WETTING

Since an important attribute of metalworking lubricants is the ability of the fluid to wet the metal surface and thus provide the desired film coverage of the surface, evaluation of this property can be useful for lubricant evaluation and control. The metal to be formed with the standard surface should be evaluated at the temperatures that will prevail during the metalworking operation.

A sample with the mill coatings likely to be present should be heated to the temperature expected to occur during the metalworking process. Subsequently, the lubricant should be applied to the surface with a dropper. The tendency of the fluid to spread or to ball up can be observed and related to its performance during actual use.

10.22 PARTICLE SIZE

The looseness of an emulsion (stability), as well as the stability of suspensions, such as graphite in water, are strongly dependent on particle size. In an emulsion, the size of the discontinuous-phase particles may vary from 0.2 to 0.5 μm, which in turn influences lubricity, stability, and sensitivity to hard water. Furthermore, during use and particularly when tramp oils are present, the particle size may slowly increase. In lubricants for forging, in which graphite is commonly used, the graphite particle size must be controlled to permit reasonable suspension times before settling occurs.

Particle size and number may be measured with a microscope capable of the size discrimination required, using an illuminated field that can be divided to provide a basis for counting the particles within a given area.

10.23 BIOLOGICAL TESTING

When the operating characteristics (such as health effects, corrosion, satisfactory welding, and ease of disposal) of the metalworking lubricant are acceptable, a very important operating characteristic remains to be tested, namely, biological stability. Bacteria, mold, and fungi may grow in the lubricant, and as a consequence of their growth, corrosion may occur, dermatitis may afflict some workers, an emulsion may be destabilized, and an otherwise satisfactory lubricant may become unsatisfactory.

Relatively simple test methods are available for measuring biological growth in a lubricant. Microbiology is generally a field of knowledge that is confusing to the engineer, who usually has not had applicable training (see Chap. 11). Several test methods useful in controlling the degradation of a lubricant and biological growth are described here. No training in microbiology is necessary for successful application of these tests.

Dip Slide Technique

This is an increasingly popular method of evaluating the growth of microbial cells. It is based on a technique of coating a plastic slide with a nutritive gel, which when dipped into the test fluid and allowed to drain, will permit microbes to grow (Fig. 10.1). The nutritive gel can be formulated to be selective for either bacteria or yeast and mold. Recently, a single slide has been developed that utilizes different media on opposite sides of the slide. On one side is an agar incorporating tetrazoleum chloride and designed to assess bacteria, and on the other side is a pink medium formulated to assess growth of yeasts and molds. Yeasts grow on the agar as pink-domed colonies. These slides are commercially available in kits.

Interpretation of bacteria and yeast infections is subject to variation due to individual judgment, as well as to genuine differences in reproducibility. However, these variations are generally of a minor order of magnitude. This on-site method provides an acceptable means of assessing microbial growth in metalworking fluids.

Oxygen Liberation Method

A new method for indirect enumeration of bacteria is based on the enzymatic conversion of a peroxide substrate to oxygen and water in a closed system. The oxygen liberated in the reaction is measured as a pressure increase via an instrument equipped with a sensitive pressure transducer. The method is rapid and gives reproducible results that are satisfactory for the control of lubricant microorganisms. Furthermore, the technique appears to result in

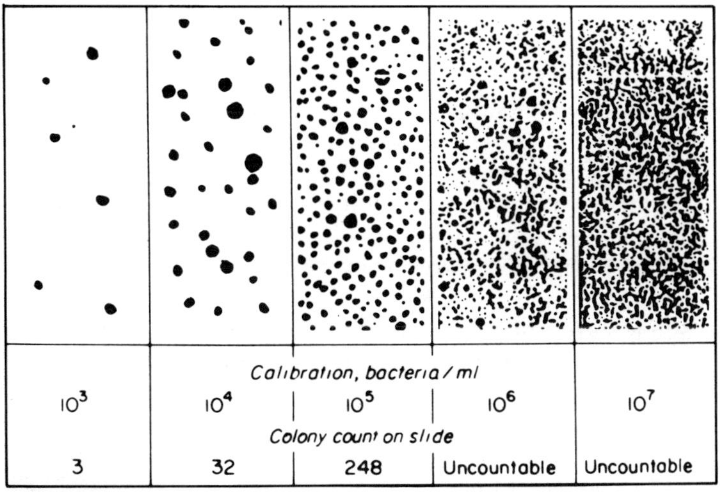

Figure 10.1 Calibration slides for estimating the extent of microbial infection in metalworking fluids. (After C. Genner and E. C. Hill, 1981. *Tribology International,* 14, 11-13. From J. A. Schey, 1983. *Tribology in Metalworking,* American Society for Metals, Metals Park, Ohio, p. 161.)

the reliable analysis of most aerobic and facultative anaerobic microorganisms contaminating metalworking fluids. The oxygen liberated by the method is measured and correlated with other standard microbial test methods to provide a standard that can be used for subsequent microbial count extrapolations.

Oxygen Concentration Test

This is a test that measures the change in dissolved oxygen in a metalworking fluid and can detect the development of aerobic bacteria. Because aerobic bacteria are usually the first to develop in most central systems reservoirs, dissolved oxygen tests should be made on a frequent, regular basis. Oxygen depletion of the fluid indicates initial degradation. A Warburg apparatus measures the oxygen consumption of living systems; the more unstable the product, the greater is the amount of oxygen utilized. This method is very rapid and produces data in 1 or 2 hr. The disadvantages are the cost of the equipment, the need for a skilled operator, and the lack of long-term knowledge of lubricant-oxygen ratio variation.

Ammonia Concentration Test

Intimate contact with metal fines in metalworking fluids, particularly those containing nitrites and/or amines, can result in the release of ammonia. Thus, periodic checking of the ammonia concentration can provide yet another means for determining unacceptable levels of metal particles as well as microbes in the fluid. It is possible to establish, by experimentation, an approximate threshold limit value for ammonia concentration in a specific control system. The amount of dissolved ammonia that can be tolerated in the fluid is determined by measuring the concentration of ammonia in the fluid without causing a discernible ammonia odor. A drop in pH with a marked increase in the level of particulate matter serves to signal the need for improved particulate control and pH level.

Frequently, an increase in ammonia is directly related to a growth in microbial content. Hence, particulate and microbial effects must be separately determined to achieve the desired quality control. An expanded-scale pH meter, equipped with a specific ion electrode called the ammonia probe, serves as a practical analytical tool for purposes of ammonia control. Very often, an increase in ammonia is related to growth in microbial content.

10.24 WATER COMPOSITION

Since water-based emulsions and solutions often contain 10, 20, and 30 times the volume of water as compared with the concentrate, the chemical composition, as well as the bacterial and fungal concentrations in the water used to dilute the concentrate for use, can be crucial. Furthermore, experience has confirmed that the characteristics of water with respect to its chemistry and bacterial or fungal content varies in an uncontrolled manner, particularly if the water supply comes from deep wells. Indeed, the composition of the water can vary from season to season. This variation, in addition to the potential concentration of chemicals in the water, influences the life and performance of the soluble or synthetic concentrate.

Determination of the elements present, particularly cations, such as magnesium, aluminum, calcium, and sodium, is important to predict the possible formation of hard-water soaps as well as the use of surfactants in complexing the cations present. A trivalent element, such as aluminum, is much more effective in complexing emulsifiers or surfactants present in emulsions or solutions than is a monovalent cation, such as sodium. Furthermore, the presence of bacteria, fungus, and mold in the freshly added water is frequently ignored as a source of contamination.

Table 10.3 Tap Water Hardness and pH for Selected U.S. Cities

City	Average hardness, ppm $CaCO_3$	Average pH
New York	33	7.0
Chicago	126	8.1
Detroit	100	7.5
Philadelphia	95	7.2
Pittsburgh	90	6.7
Los Angeles	110	8.4
Cleveland	118	7.4
St. Louis	98	9.1
Milwaukee	124	7.6
Youngstown	78	10.4
Cincinnati	113	8.4
Minneapolis–St. Paul	78	8.4
Allentown–Bethlehem	180	7.4
Indianapolis	290	7.5
Kansas City, MO	110	10.0
Hartford	12	6.5
Flint	84	9.8
Seattle	28	7.2
Houston	55	7.8
Birmingham	51	8.4
Atlanta	25	8.7
Grand Rapids	129	8.0
Trenton	48	7.6
Peoria	346	7.2

Also significant is the effect of increasing hardness as water is added progressively to a recirculating system to maintain the volume and concentration during use. The constant replenishment of water, in due course, greatly increases the hardness of the water in the system, far beyond that measured in the water being added. Thus, a knowledge of water chemistry and contaminants, both prior to use and during manufacturing operations, is vital in maintaining the optimum characteristics built into the metalworking fluid.

The total hardness, that is, the solution of calcium and magnesium in water, is generally expressed in terms of the calcium carbonate ($CaCO_3$) equivalent and may be titrated directly using appropriate indicators. Other cations, such as iron, may be present in the water and affect the emulsion or solution and also may be titrated. Determination of sulfate and chloride anions similarly is useful in evaluating the acidity of a metalworking fluid.

The hardness of water is critical with respect to the inherent stability to be expected of the lubricant. Typical hardness and the pH of tap water in selected U.S. cities are given in Table 10.3. These characteristics of water must be considered and adjusted when necessary in order to maintain the stability as well as operating characteristics of the metalworking fluid.

10.25 CONDUCTIVITY

A simple conductivity meter may be used to evaluate the intrinsic electrical conductivity of a water-based fluid. Regular readings obtained on the meter can serve as a monitoring check on the buildup of salts or metallic impurities in the fluid, thus forewarning of the necessity for cleanup or disposal.

The tests described in this chapter are only illustrative of the many complex, as well as simple, test procedures that are useful for the control of metalworking lubricants. The importance of small-scale pilot operations, using the lubricant as the final arbiter of performance, should be emphasized prior to full-scale use. Nevertheless, control methods to maintain performance are imperative.

BIBLIOGRAPHY

ASTM Standards, Section 5, Volumes 05.01; 05.02; 05.03 and 05.04.

Beckert, G. J. P., 1982. Knowing the true concentration is the key to longer cutting fluid life, in *Lubrication in Metalworking,* Third International Colloquium, Esslingen, Germany, PP. 105.1–105.12.

Bennett, E. O., 1974. The biological testing of cutting fluids, *Lubrication Engineering,* 30(3), 128–135.

Dorsey, G. A., 1979. An on-line emulsion coolant monitoring procedure, in *Lubrication Challenges in Metalworking and Processing,* Proceedings of the Second International Conference, IIT Research Institute, Chicago, pp. 171–175.

Gannon, J. E., and E. O. Bennett, 1981. A rapid technique for determining microbial loads in metalworking fluids, *Tribology International,* February, 3–6.

Genner, C., and E. C. Hill, 1981. Evaluation of the dip slide technique for cutting oils, *Tribology International,* February, 11–13.

Jain, S., 1982. Water-based hot forging lubricant: The ideal non-pollutant, in *Lubrication in Metalworking,* Third International Colloquium, Esslingen, Germany, pp. 29.1-29.5.

Lockwood, F. E., and I. J. Matienzo, 1982. Contaminant effects in aluminum hot rolling emulsions, in *Lubrication in Metal Working,* Third International Colloquium, Esslingen, Germany, pp. 50.1-50.6.

Russ, G. A., 1980. Coolant control of large central systems, *Lubrication Engineering,* 36 (January), 21-24.

Smith, M. D., and J. E. Lieser, 1973. Laboratory evaluation and control of metalworking fluids, *Lubrication Engineering,* 29 (July), 315-319.

11

Toxicological and Biological Aspects of Metalworking Lubrication

Some of the possible health effects that may be experienced by some workers exposed to metalworking lubricants are discussed in the first part of this chapter. We will attempt to clarify some of the relevant factors and discuss practices that will minimize possible ill effects on human health. The second part of this chapter is devoted to a discussion of the growth of microorganisms in metalworking lubricants. Among various other effects, degradation of lubricants may be a direct consequence of microorganism infestation of the lubricant.

Much remains to be learned about both the toxicity and the microbiology of metalworking lubricants. The protection of those who come in contact with lubricants regularly, and on a long term basis, has received more attention as public awareness of the importance of the working environment grows. Because of the state of knowledge and fear of the unknown, appropriate health measures are often complicated by a lack of information. In the specific case of exposure to metalworking lubricants, the complexity of that environment makes appropriate analysis of possible toxic effects difficult.

11.1 TOXICOLOGICAL ASPECTS OF METALWORKING LUBRICATION

The air and any fluids or surfaces that come into contact with humans can carry with them substances that may irritate or infect the skin and, if they are potent enough, alter human cellular structure. Everyday living carries with it the potential for a toxic reaction to the environment. Toxic effects may be triggered by the sun, sugar and/or salt in food, plant poisoning (poison ivy), and a large number of other contacts. Furthermore, not only are these hazards always present in the particular environment in which we find ourselves, but because of the complexity of both the environment and our own biochemistry, definitive studies on toxicological hazards are often both time consuming and very expensive. Furthermore, the results of such investigations can be inconclusive.

The amount of exposure to metalworking lubricants, as well as which lubricant components influence the incidence of human ill effects, is related in a complex manner to other phenomena with which the body copes (viral infections) and various other environmental effects (sunburn). Fortunately, more frequently than not, most humans in a given environment are totally unaffected by possible damaging biological effects. However, some individuals may incur toxicological damage that is not observed in the general population. Thus, in considering the toxicity and biological aspects of metalworking lubricants, the complexity of the questions raised should be appreciated. Furthermore, an understanding of our present incomplete state of knowledge is important in attempting to resolve any relevant health questions.

11.1.1 Skin Disorders and Other Possible Health Effects

Skin disorders (dermatoses) may result from frequent contact with chemicals contained in lubricants. Most often, this frequency is related to the use of metal cutting lubricants, since machine operators tend to be in more frequent contact with the lubricants over a long period of time.

The skin disorders resulting from prolonged contact with metal cutting lubricants are of four major types:

1. Contact dermatitis
2. Folliculitis and acne
3. Pigmentary changes
4. Benign and malignant tumors

Contact Dermatitis

Chemical reactions at the surface of the skin are most frequently thought to be the cause of dermatitis. Water, solvents, emulsifiers, soaps, and detergents may defat the skin through the removal of surface lipids. Thus, they may contribute to the possible occurrence of contact dermatitis. Furthermore, excessive alkalinity (pH 9 and above) may contribute to the occurrence of dermatitis.

Folliculitis

If prolonged contact is maintained with oil-based lubricants, the oil may block the hair follicles, blackheads may develop, and inflammation may flolow. Bacteria present at the surface may also contribute to folliculitis as well as oil boils. Folliculitis occurs most frequently on the backs of hands, forearms, thighs, face, and back of the neck.

Pigmentary Changes

Although sunburn has been implicated as a principal cause of skin darkening and thickening, these changes have also been reported associated with folliculitis.

Benign and Malignant Tumors

Polycyclic aromatic hydrocarbons present in base oils used in formulating metalworking lubricants may cause scaly keratoses as well as benign and malignant tumors. Solvent refining of these oils, which removes most of these aromatic compounds, has been adopted as a practice for oils used in metalworking fluid formulations.

Some additives used have been identified as having a greater potential for some degree of toxicity to humans. These additives are classified as follows.

1. Extreme-pressure additives: chlorine, phosphorus, and sulfur compounds
2. Bactericides: formaldehyde and formaldehyde donors, halogenated salicylanilides, mercaptobenzothiazoles
3. Corrosion inhibitors: hydroxylamines, inorganic and organic nitrites and nitrates, zinc dithiophosphate, and petroleum sulfonates
4. Dyes and coloring agents: fluorescein, petroleum azo dyes, and paraphenylenediamine
5. Water conditioners and antiwear agents: phosphates and borates
6. Emulsifiers and detergents: soaps, petroleum sulfonates, and calcium alkyl salicylate
7. Odorants: mint oil, pine oils (terpenes), and tars
8. Antioxidants: amines (diphenolamine), hydroxy compounds, alkyl sulfides and disulfides, zinc diethyldithiophosphate

Suspicion of possible contribution to toxic effects may also extend to many more chemicals with which humans are daily in contact. Nevertheless, it is prudent to reduce the contact time with fluids that may have one or more of these constituents. Direct evidence of any deleterious toxic effects is, at present, inconclusive.

Minute metal slivers that accumulate in metalworking fluids may abrade the skin. Slivers that are often collected on plant towels used by the workers can scratch the skin surface. This damage to the skin, resulting either from the chemistry or physical constituents in the metalworking fluid, provides loci for initiation of dermatitis.

The air in the plant environment also provides a carrier for inhalation of oil that has become vaporized. Additives may also be carried by vaporized oil and/or water, as well as by microorganisms present in the atmosphere. Cases have been reported that have been attributed to air-carried toxic materials, resulting in pneumonia, sore throats, and laryngitis.

In the case of cutting fluids, one of the major classes of metalworking lubricants, dissolved metals that leach from the metals being processed have been shown to cause sensitization of the skin. Buildup of metals in the fluids can be expected to increase as the fluids are used over a long period of time to reduce their initial cost. Nickel, chromium, and cobalt are the three most common skin sensitizers. Individuals who are allergic to cobalt often exhibit allergies to nickel and chromium if these are present in the lubricant. Sensitization of the skin may or may not lead to dermatitis. However, the potential must be kept in mind by those responsible for preventing dermatitis outbreaks in the plant.

11.1.2 Biological Screening Procedures

Because of the low order of toxicity of most metalworking lubricants and the complex nature of the environment, measurement of the toxic effects that may result from chronic exposure is very difficult.

Recent developments of short-term and in vivo toxicological testing methods have reduced the cost and time required for evaluating materials that are possibly carcinogenic or mutagenic. Mutagenicity (the ability to alter chromosones) can result in the growth of defective cells, as well as cause changes in cellular function. Cells undergoing rapid and unregulated division and growth can lead to the development of a tumor. Thus, the ability to detect a mutagenic event provides a screening procedure for identification of substances that are potentially carcinogenic.

The patch test (small patches soaked in the suspect material are taped to the forearm for a time not exceeding 48 hr and removed, and the skin inspected for irritation or the onset of dermatitis) is a commonly used procedure for carrying out comparative evaluation of various fluids as they may affect a specific person. The Ames test (effect of test materials on bacteria) has reduced the cost and time for determining the potential for mutagenic effects.

Several specially selected strains of bacteria are exposed to the test material, and the number of mutant bacteria is measured. This result is then used as a measure of carcinogenicity.

The effective exposure of a potentially toxic substance depends upon the length of time of exposure to the substance, ease of entry into the body, the rate of excretion from the body, and/or the rate of accumulation within tissues. If patch tests or the Ames test indicates the possibility of toxicological effects, then animal testing, primarily on rabbits, may be performed using the suspect material. The battery of tests normally carried out using rabbits includes acute inhalation toxicity, acute eye irritation, and primary skin irritation. Acute oral and inhalation toxicity tests are carried out using rats.

It should be emphasized that animal testing, the patch test, and the Ames test are indicative of possible mutagenic or carcinogenic effects associated with the material being tested. Positive proof of cancer-causing effects in humans is only discovered through long-term exposure of a human population.

11.1.3 Prevention and Control of Dermatitis

Prescreening of workers can be an effective means of eliminating the incidence of dermatitis. Questions regarding previous skin disorders and medical history can be very useful in this regard. Generally, the lighter the skin of the worker, the greater is the hyperreactive response to irritants. The darker-pigmented skin tends to be less sensitive to potential irritants. Lighter-skinned individuals with inherently oilier skin or more profuse hair follicles appear to be less susceptible to skin irritation. Individuals with red hair tend to be more acutely susceptible. Workers with severe acne are also especially susceptible.

Generally dermatitis occurs during the first year of employment, if it occurs at all. Outbreaks appear to be associated with the winter months because of low humidity, which tends to produce skin fissures.

Elimination of dust and vapor in the air and minimization of physical contact with the chemicals in the metalworking lubricant, as well as proper attention to individual cleanliness, all contribute to minimizing the incidence of dermatitis.

An organized program designed to reduce the onset of dermatitis can be very effective. A recommended program in a plant might consist of the following.

1. Work clothing should be clean. Splash guards should be fastened to the machines, and the worker should be protected by wearing impervious armlets and aprons.

2. Before starting work, before eating, and before leaving the plant, workers should be encouraged to wash with a mild soap. Soaps that contain harsh abrasives should be avoided. After washing, an oily hand cream should be used to maintain the skin oil level.
3. Solvents of all types, including gasoline, kerosene, and similar compounds, should not be used to clean the skin.
4. To eliminate spreading, towels or wipe cloths should not be interchanged among workers. Wiping materials should be washed frequently or discarded after use.
5. Care should be taken to prevent clothing from becoming soaked with the lubricant.
6. Any human discharge into the metalworking lubricant or use of a reservoir as a garbage pit should be prevented.
7. Every cut and scratch should be reported and taken care of as soon as possible.
8. Lubricants should be filtered to remove fines and other fine particles from the fluid.
9. The lubricant chemistry, pH, and biocidal content must be controlled within recommended limits. Only approved biocides should be used and even then with care.
10. In the interest of good housekeeping, the equipment and the surroundings should be kept clean and reservoirs should be cleaned regularly to remove any accumulation.
11. A continuing program of education stressing the importance of personal hygiene should be carried out.

Determination of the causes of dermatitis is difficult due to the complex environment in the plant. Personal health at a particular time may contribute to the occurrence of dermatitis. Paper toweling, soaps, personal cleanliness habits, eating habits, humidity, skin punctures or abrasions, suntan lotion, and shaving cream are only some of the potential ingredients of a dermatitis-producing environment.

Above all, personal and environmental cleanliness must be basic to a healthy work environment. In addition, the metalworking lubricant must be formulated, controlled, applied, and maintained according to good practice if the health of the workers is to be protected.

11.2 MICROBIOLOGY OF METALWORKING FLUIDS

The growth of microbes (bacteria, molds, and yeast) occurs in metalworking fluids when water is present. The emulsions or solutions primarily used for metalworking are most commonly subject to microbial attack. However, microbes may also grow in pastes, oil, and gels when water is present, either as a contaminant or as a component. In this section we will discuss and describe the microbes of importance, their effects, and control.

Microorganisms are extremely small cells that have a capability for reproducing very rapidly in a wide variety of environments. The microorganisms of interest in metalworking lubricants are classified as either bacteria or fungi. In general, bacterial cells are much smaller than fungal cells. Fungi may be broken down into two subcategories, called molds and yeasts. Under the microscope, molds appear to be rough and fuzzy and yeasts look smooth and glisten. The molds look like a tangled mass of threads under the microscope; yeasts appear as individual cells.

The microorganisms that feed upon the solutions and emulsions important in metalworking can grow and be sustained by a wide range of nutrients (Fig. 11.1). Microbes feeding on an oil emulsion can double their body weight and

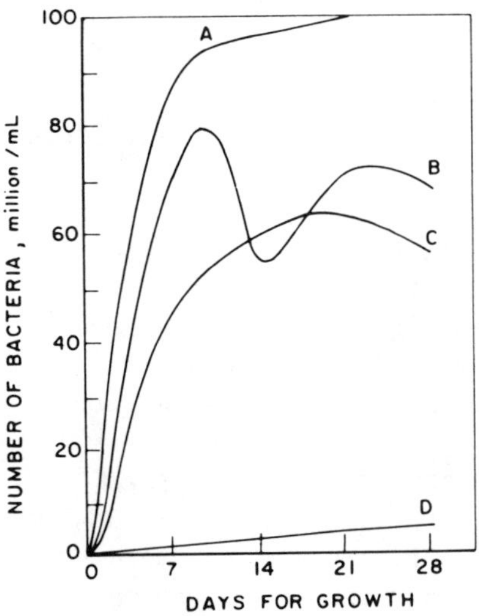

Figure 11.1 Rate of bacterial growth in water-based lubricants as a function of available nutrients: (A) lard oil, (B) sulfurized oil, (C) paraffin oil, (D) synthetic coolant. All dilutions are 1:25. (*Source:* Master Chemical Corp. From J. A. Schey, 1983. *Tribology in Metalworking,* American Society for Metals, Metals Park, Ohio, p. 158.)

then divide into two every 20 min. Obviously, the emulsion present after 1 day or 1 week of such growth at the expense of the constituents of the emulsion has fundamentally altered characteristics. The exact microbiology of any metalworking system depends upon the rate of aeration, if any, the temperature, the pH, the rate of makeup, the availability of nutrient elements and their composition, and the presence of growth inhibitors.

11.2.1 Microbial Effects of Lubricant Chemistry

There is some evidence that the microorganisms present in our general surroundings also may be present in metalworking fluids and can aggravate dermatitis. Furthermore, these microorganisms or microbes present as bacteria or fungi can significantly change the chemistry of the lubricant, thus modifying its chemical and physical behavior.

The chemical changes consist of oxidation, which results in the removal of side chains from complex molecules, the opening of aromatic rings, reduction of carbon chain lengths, and unsaturation of saturated bonds. There is usually a reduction in pH, which can be masked if ammonia is liberated due to the degradation of nitrogen-containing additives in the lubricant. Additives containing phosphorus, sulfur, and nitrogen may be degraded and may reappear as inherent constituents of the growing microbes. Sulfur often reappears as hydrogen sulfide gas, so obnoxious in the workplace and identified as "Monday morning stink."

The chemical changes affect lubricating properties and characteristics. The additives present in the lubricant for the purpose of preventing corrosion of the workpiece and tooling lose their effectiveness as they degrade. Frequently, corrosion begins where no corrosion has occurred before. In the case of emulsions, changes in the chemistry of the emulsifying agents increase particle size and, if carried far enough, can result in separation of the emulsion, thus changing its physical characteristics.

It is significant that microbes present in metalworking fluids often start growth elsewhere in the environment. The air in the plant, the water used for makeup of lubricants containing water, or, in the case of contaminated lubricants, contamination with water and human contact, as well as previous buildup of biological material on equipment, all provide microbe sources.

11.2.2 Some Significant Microbes

The microorganisms present in metalworking lubricants and most particularly associated with lubricants containing water are of three main types:

1. Aerobic bacteria, primarily the *Pseudomonas* group and, less important, coliforms
2. Anaerobic bacteria that are sulfate reducers, particularly *Desulfovibrio desulfuricans*
3. Fungi, principally *Fusarium* and *Cephalosporium;* the species *Candida* (yeast) is also found

The anaerobic bacteria, which grow in the absence of air (oxygen), are the bacteria responsible for breaking down organic and inorganic sulfur-containing compounds. Aeration of a reservoir containing these bacteria can often effectively kill these bacteria. The aerobic bacteria are the most aggressive and prevalent in degrading the lubricant and grow in the presence of air (oxygen). They are the principal cause of biological deterioration of the metalworking emulsions.

Fungi develop preferentially in most metalworking water-based solutions. The microorganisms that cause problems are typically either molds or yeasts. Although they may alter the chemistry and performance of the lubricant, these effects are generally long range. Through periodic additions of the concentrate used to make up the solution, degradation of the fluid is compensated for. The predominant problem is mechanical, due to the buildup of mold that may block filters, cause buildup on tooling and equipment, and, because of their obnoxious appearance, be objectionable to the worker.

11.2.3 Control of Microbes

Microorganisms may be brought under control by active and continuous attention to cleanliness of the workplace. Efforts in this regard are the most cost effective.

Selection of a metalworking lubricant that is inherently resistant to microorganism growth minimizes the occurrence of microbial problems. Some constituents of a lubricant are more prone to microbial degradation than others, and if they are not present, the life of the lubricant can be extended. The development of a microbial-resistant lubricant is, however, not an unmixed blessing. In a sense, degradation of lubricants to be disposed of depends upon the ease with which they may be biologically degraded.

Numerous methods for controlling microbial infestation, as an alternative to biocidal treatment, have been investigated, such as the application of radiation of various forms (ultraviolet light or γ irradiation), pasteurization carried out on a batch or continuous basis as well as filtration, and centrifugation. Among these methods, pasteurization is the most cost effective in application and will probably grow in use, particularly when large central systems containing solutions or emulsions are in use.

Table 11.1 Some Biocides that are EPA Registered for Use in Metalworking Fluids

Manufacturer	Trade name and active ingredients	Recommended dose (% of formulated biocide)	EPA reg. no.	Comments
Buckman Laboratories, Inc.	Busan 85: potassium dimethyldithiocarbamate 50%	0.03–0.2	1448–70	Compatible with most concentrates (exceptions usually oil-water emulsion)
The Dow Chemical Co.	Dowicil 75: 1-(3-chloroallyl)-3, 4,7-triaza-*l*-azoniaadamantane chloride (67.5% sodium bicarbonate inert stabilizer)	0.015–0.2	464–403	Not recommended for fluid concentrates, usually added on site; hygroscopic powder
	Dowcide 1: *O*-phenylphenol 98%	0.1–0.15	464–70	Compatible with most concentrates; restrictions on discharge levels of phenols – inhibits use
	Dowcide A: sodium *O*-phenylphenate 97%	0.1–0.15	464–78	Compatible with most concentrates; restrictions on discharge levels of phenols – inhibits use
	XD-8254 DBNPA: 2,2-dibromo-3-nitrilo-propionamide 10%	0.02–0.04 daily	464–500AA	Cannot be used in concentrates; unstable in alkaline pH; usually added continually – degrades easily

Givaudan Corporation	Givgard DXN: 6-acetoxy-2,4-dimethyl-*m*-dioxane 100%	0.05-0.02	824-7	Incompatible with most concentrates
ICI United States Inc.	Proxel CRL: 1,2-benzisothiazolin-3-one 30-35%	0.001-0.06	10182-3	Stable in many concentrates; not aggressively being marketed by manufacturer for metalworking fluids at this time
IMC Chemical Group, Inc.	Bioban P-1487: 4-(2-nitrobutyl)-morpholine 70%; 4,4′-(2-ethyl-2-nitrotrimethylene) dimorpholine 20%	0.01-0.3	271-30	Good oil solubility; stable in oil-water emulsion concentrates
	Tris Nitro: tris (hydroxymethyl) nitromethane; aqueous 50%, powder 100%	0.2 aqueous 0.1 powder	271-26 271-18	Poorly compatible with most concentrates; recommended for tankside use
Lehn & Fink Industrial Products, Div. of Sterling Drug Inc.	Grotan HD2: 2-chloro-*N* (hydroxymethyl) acetamide 39%; sodium tetraborate 41%; potassium iodide 0.39%	0.1-0.15	10,000-3	Incompatible with concentrates; recommended tankside; soluble oil needs higher dose
	Grotan BK[a]: hexahydro-1,3,5-tris (2-hydroxyethyl)-*s*-trizane 78%	0.15	10,000-1	Stable in many concentrates: less effective against fungi at low dose levels

Table 11.1 (Continued)

Manufacturer	Trade name and active ingredients	Recommended dose (% of formulated biocide)	EPA reg. no.	Comments
Olin Corp.	Zinc Omadine: zinc 2-pyridinethiol-1-oxide			Not soluble in concentrates
	powder 95%;	0.0079	1258–840	
	aqueous dispersion 48%	0.015	1258–841	
	Sodium Omadine: sodium 2-pyridinethiol-1-oxide			Soluble in most concentrates except with difficulty in oil-water emulsion concentrates; more effective against fungi and yeasts than bacteria
	powder 90%;	0.005	1258–842	
	aqueous solution 40%	0.0115	1258–843	
	Triadine 10: hexahydro-1,3,5-tris (2-hydroxyethyl)-*s*-triazine 63.6%; sodium 2-pyridinethiol-1-oxide 6.4%	0.07 in synthetic fluids; 0.1 in oil-containing fluids	1258–990	Used in oil-water emulsion concentrates with care; compatible with most other concentrates
Onyx-Millmaster	Onyxide 200: hexahydro-1,3,5-tris (2-hydroxyethyl)-*s*-triazine 78%	0.15	1839–59	Stable in many concentrates; less effective against fungi at low dose levels

Rohm and Haas Co.	Kathon 886 MW: 5-chloro-2-methyl-4 isothiazolin-3-one 8.6%; 2-methyl-4-isothiazolin-3-one 2.6%	0.0025–0.0125	707–129	Stable in some concentrates; not particularly in oil-water emulsion concentrates – possibly inactivated by some amines and sulfides
R. T. Vanderbilt Co., Inc.	Vancide TH: hexahydro-1,3,5-triethyl-*s*-triazine 95%	0.05–0.1	1965–55	Good oil and water solubility; stable in most concentrates; strong amine odor
	Vancide 51: sodium dimethyl-dithiocarbamate 27.6% sodium 2-mercaptobenzothiazole 2.4%	4% in water as bactericide; 2% in water as fungicide	1965–8	Stable in some concentrates but recommended at-use dilution by user

[a]This product is sold as "Grotan" in the United States.

The most common and widely used method of control consists of the application of biocides to the lubricant, either by the supplier of the concentrate (in the concentrate) or by the user to the lubricant reservoir. It is the latter addition at tankside that has proven to be the most effective.

The biocides of commercial importance vary broadly in chemistry and may attack different strains of bacteria and fungi with varying degrees of effectiveness (Table 11.1). Selection of a biocide is thus important and should involve an understanding of whether the particular biocide is effective against the bacteria and fungi apt to be present in the particular environment. Furthermore, will the biocide be chemically compatible with the lubricant, or will it chemically react with some components and thus degrade the lubricant, as well as end biocidal activity?

Establishment of a safe procedure for handling tankside additions of the biocide should be established, as well as any special conditions involved in disposal of the treated metalworking lubricant. No biocide should be used that has not been approved by the Environmental Protection Agency. Such approval involves extensive testing to determine possible toxicological effects. Finally, the cost effectiveness of one biocide, as compared with another, should be determined. As is the case in the selection of the metalworking lubricant, the highest initial cost should not predetermine the cost effectiveness of the additive.

Assuming that steps have been taken to organize a clean workplace and that value analysis judgments have been made about the selection of the lubricant as well as the biocide, there remains the need for ongoing monitoring of the microorganisms that may enter and grow in the lubricating system.

11.2.4 Monitoring

Assuming that systematic procedures have been established for ensuring cleanliness, periodic testing of the microbiology of the lubricant is desirable to avoid the processing problems resulting from biological degradation of the lubricant. Some of these direct methods are described in Chap. 10. Indirectly, tests that may be carried out on a regular basis to measure emulsion stability, pH, corrosion propensity, lubricant concentration, color, viscosity, and particulate contamination can also provide useful information upon which to make judgments about the presence or absence of lubricant deterioration due to biological infestation. Unfortunately, changes in these criteria may also result from other operating occurrences. Nevertheless, when a system has been systematically placed under control, changes in a measured characteristic can, with experience, be related to the source of the problem.

Some specific analytical techniques that may be used to directly evaluate the presence and concentration of microorganisms are described in Chap. 10. Dip slide techniques and measurement of lubricant oxygen content (a technique for measuring the oxygen released as a result of microbiological infection), as well as standard plate count methods, are effective tools for use in the control of microorganisms in lubricants.

BIBLIOGRAPHY

Ames, B. N., J. McCann, and E. Yamasaki, 1975. Methods for detecting carcinogens and mutagens with the salmonella mammalian microsome mutagenicity test, *Mutation Research,* 31, 347-364.

Bartz, W. J., and J. Wolff, eds., 1982. *Lubrication in Metal Working,* Technische Akademie Esslinger, Germany, 80.1-85.4.

Bennett, E. O., 1972. The biology of metalworking fluids, *Lubrication Engineering,* 28(7), 237-247.

Bourne, L. B., 1969. *Skin Disease from Oil: Causes and Prevention,* Mason House, London.

Gilbert, R. J., and D. W. Lovelock, eds., 1975. *Microbial Aspects of the Biodeterioration of Materials,* Academic Press, London.

Good Plant Practices for Workers Using Cutting Fluids, 1978. American Petroleum Institute, Washington, D.C.

Hall, E., 1967. How bacteria damage lubricants, *New Scientist,* 17 (August).

Olishifski, J. B., ed., 1979. *Fundamentals of Industrial Hygene,* 2nd ed., National Safety Council, Chicago.

Plunkett, E. R., 1976. *Handbook of Industrial Toxicology,* Chemical Publishing Co., New York.

Rossmore, H. W., 1980. Antimicrobial agents for water-based metalworking fluids, American Occupational Medical Association, *Proceedings of the Sixty-Fifth Annual Meeting,* Detroit, Michigan, April 24, 199-219.

Taylor, J. S., 1979. Dermitatis associated with metalworking fluids, in *Lubrication Challenges in Metalworking and Processing,* IIT Research Institute, Chicago, Illinois, pp. 239-245.

12
Lubricant Clarification, Recycling, and Disposal

A number of factors that influence the metalworking process are relevant when considering lubricant clarification, recycling, and disposal. Specifically, the cost of oil- and petrochemical-derived additives has increased rapidly due to real and imagined shortages, as well as profound changes in the effective control of crude oil. Therefore, it becomes more and more desirable to conserve and recycle lubricants. At the same time, laws and regulations established to protect the environment have influenced the chemistry as well as the cost of lubricant disposal. Furthermore, laws and regulations mandating practices designed to promote plant safety and to safeguard workers' health have changed lubricant application and control methods and their chemistry.

Extension of the useful life of lubricants depends upon a knowledge of the contaminants and their characteristics and control of lubricant characteristics and quality. Decisions on the choice of clarification, recycling, and disposal technology depend upon specific metalworking operating conditions.

In this chapter the changes in chemistry and contamination that may take place in the lubricant during metalworking operations are reviewed. Clarification, recycling, and disposal technology is then described and related to metalworking lubricants.

Selection of appropriate clarification, recycling, or disposal procedures depends upon an analysis of the lubricant characteristics as well as the presence of possible contaminants. If the lubricant is oil based, some of the chemical components of the lubricant may, with time, precipitate and circulate within the fluid, or they may be deposited on surfaces wetted by the fluid. If solid lubricants are used in suspension, these too may be deposited. If emulsions

Table 12.1 Particle Size Scale

References	Micrometers	Particles to be filtered
	100	Machining particles
Human hair		
	80	
	60	
	40	
Particle visible to unaided eye		
	20	Grinding particles
	10	
		← Emulsion oil droplet size range
	8	
Red blood cell	6	
Optical microscope capability →	5	Glass particles
	4	
	3	Honing particles
Yeast cells	2	
	1	Preformed emulsion droplet size range
	.8	←
	.6	Precipitates and scum
	.5	
	.4	
	.3	
Bacteria	.2	
	.1	True solutions
Viruses		

Source: R. H. Brandt, 1982. Clean coolant is rewarding, in *Industrial Fluids Controls, Concerns and Costs,* M. Opachak, ed., Society of Manufacturing Engineers, p. 59.

(soluble oils) are used, the stability of the emulsion can be reduced and the particle size of the discontinuous phase can increase. These changes influence the choice of the clarification, recycling, or disposal process and equipment. If solutions (synthetics) are used, changes in pH may cause precipitation. Pressure and temperature effects can then result in agglomeration of the chemical components present in the solution. Finally, in many metalworking lubricants there is a potential for biological degradation and resulting organic particulate contamination, which must be removed to maximize the effectiveness of the lubricant.

Metalworking lubricants, particularly those to be recycled, may contain a wide diversity of solid or liquid contaminants, the chemistry, size, geometry, and concentration of which should be evaluated in terms of their impact on the clarification, recycling, and disposal technology (Table 12.1). Particles, present in the fluid, are generated from many possible sources. Frequently, the contaminant is present as a result of random events, such as the introduction of cleaning fluids into the lubricant stream, sweeping of oil-absorbent materials from the floors into flumes, disposal of food into reservoirs, and various other random contamination sources. Obviously, these random occurrences are difficult to take into account on a systematic basis, but their occurrence should be kept in mind as the process and equipment decision for clarification, recycling, and/or disposal is made.

Contamination due to particulates from the following sources are more readily controlled.

1. Tooling particles. In grinding, aluminum oxide, silicon carbide, or diamond grit, as well as the bonding material of the wheel (resin or vitrified), may all contaminate the lubricating fluid. The organic bonding materials may, if present in the fluid and not removed, deposit as gummy coatings. The inorganic materials vary in size and shape. Materials and their characteristics influence the selection of the clarification equipment. In various other metalworking operations, the tools and dies may wear or fracture and the debris will be carried by the metalworking fluid.
2. Workpiece particles. The chemistry and structure of the workpiece material influence the character of the particulates produced as a result of the metalworking process. In the case of cast iron, for instance, iron oxide, silicon oxide, and graphite particles may be present in the fluid. In the case of aluminum and its alloys, aluminum oxide is often present in the recirculating lubricant. All the metals commonly used in metalworking operations also serve as a source of particulates.
3. Water. The presence of such cations as calcium and magnesium promotes the formation of insoluble precipitates, particularly when soaps are a

component of the lubricant. Furthermore, various metal sulfates and chlorides may be totally or partially dissolved in the lubricant emulsion or solution. Either during recirculation of the lubricant or in the process of clarification and recycling, crystalline participates may be produced since their solubility may be exceeded. Knowledge of these in-process changes in the lubricant can markedly improve the basis for deciding on a particular clarification equipment and process design of a recycling system.

4. Biological particulates. The growth of biological organisms, such as bacteria and fungi, if unchecked, not only degrades the chemistry of the metalworking lubricant to the point where precipitation of unwanted by-products occurs, but also may, particularly in the case of fungi, such as mold, significantly block filters. A completely different effect of biological degradation may be reduced corrosion protection, possibly resulting in the corrosion of piping and other components of the lubrication system, which can generate solid corrosion particulate by-products. Furthermore, if the lubricant reservoir or flume is constructed of concrete, microbes can attack the concrete, thus providing a source of lime and/or calcium carbonate to further contaminate the lubricant.

Thus, lubricant degradation and particulate contamination often interfere with and can profoundly affect the cost effectiveness of a given lubricant clarification, recycling, and disposal system. An understanding of the chemical and biological processes that may occur and the resulting effects are thus essential.

12.1 CLARIFICATION

As suggested, many factors encourage extending the effective life of a given metalworking lubricant. A significant contribution to extending the life of a given lubricating system is the installation of appropriate clarification equipment. A number of different technologies are briefly described in the sections that follow (see Fig. 12.1).

12.1.1 Settling

Particulates in many metalworking lubricant systems are often adequately removed by installation of a simple gravity-settling tank (Fig. 12.1a). To facilitate removal of the settled debris, fines, and mold, a scraper chain is often installed that serves to remove the buildup from the bottom of the tank. Separation of the tank into separate sections by partitioning walls over which the fluid flows increases the effectiveness of the settling system by progres-

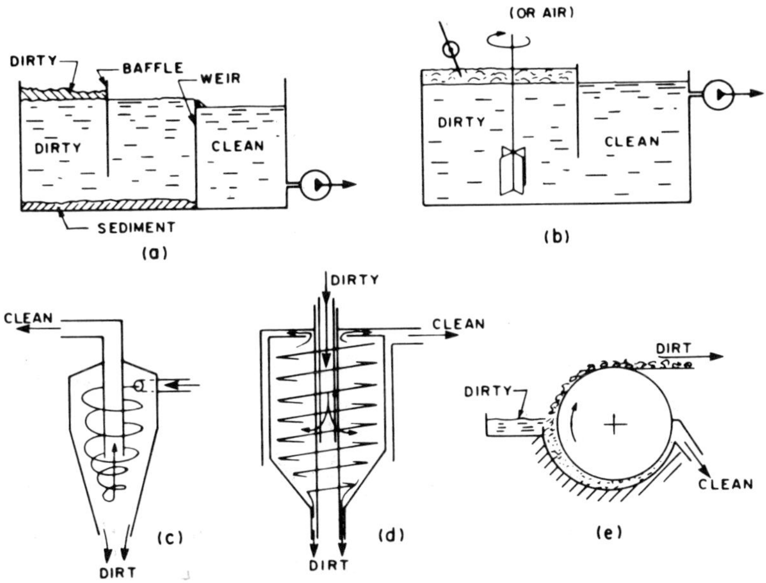

Figure 12.1 Methods of removal of fines and contaminants from metalworking fluids: (a) settling tank, (b) flotation, (c) hydrocyclone, (d) centrifuge, (e) magnetic drum. (After J. A. Schey, 1983. *Tribology in Metalworking*, American Society for Metals, Metals Park, Ohio, p. 182.)

sively reducing the contaminant that may flow from one section to the other. Furthermore, a baffle may be installed at the top of the tank to prevent the fluid from flowing over the top of the wall, thus providing a barrier to tramp oil.

Metalworking fluids that are commonly treated by such gravity-settling devices should be formulated by appropriate chemical modifications to reduce suspension of particulates and minimize foaming. These compositional changes are particularly important in water-based fluids. In straight or compounded oils, gravity-settling systems are less complex.

Gravity-settling systems, because of their low cost and simplicity, are often used as the first stage in a system that may include cyclone separators, filtration devices, and/or a centrifuge. As a result, the operating cost of the more sophisticated equipment is significantly reduced and the cost effectiveness of the total system is frequently improved.

12.1.2 Skimming

Removal of tramp oil from the surface of a gravity-settling tank is simple. It involves the use of a rotating disk or belt wetted by the oil. The belts may be made from stainless steel, neoprene, polypropylene, or other materials. The tramp oil preferentially wets the belt and is continuously removed from the surface of the lubricant.

To facilitate the separation of oil that may have become partially emulsified in the metalworking fluid, aeration units are frequently used. The aeration devices used to facilitate tramp oil removal generate air bubbles to which the tramp oil adheres and thus floats to the surface, bringing with it not only the tramp oil but also fine precipitated particles of metal oxides and other impurities.

A combination of settling tank, skimmers, and aeration devices (Fig. 12.1b) will remove coarse particles by settling and fine particles and tramp oil by skimming. This relatively inexpensive combination often provides sufficient "clean" lubricant acceptable for repeated cycling.

12.1.3 Cyclone Separators

Cyclone separators magnify the effect of gravity in achieving particulate separation through the use of centrifugal force. The contaminated fluid is centrifugally accelerated by a pump through a cone with decreasing diameter (Fig. 12.1c). The centrifugal force exerted on the fluid drives the suspended particles to the inner wall of the cone. These particles then build up due to gravity, then fall and are discharged from the bottom. The back pressure resulting from the geometric configuration of the cone and the discharge nozzle causes the cleaned fluid to discharge from the top. The hydrocyclone may be used most effectively subsequent to the use of settling tanks to remove residual contaminants. The equipment is effective only on relatively light viscosity fluids of approximately 60–100 SUS at 100°F (38°C).

If tramp oil is present, the hydrocyclone tends to increase its emulsification. This emulsification can lead to problems of fluid breakdown, bacterial growth, corrosion, and staining. Furthermore, unless care is taken, air may be sucked into the hydrocyclone through the bottom exhaust port, thus creating foaming problems and stable air-in-water mixtures, which may seriously interfere with fluid clarification.

When contamination by tramp fluids is minimal, hydrocyclones nevertheless provide a relatively simple maintenance-free device for removal of solid contaminants from a low-viscosity metalworking lubricant.

12.1.4 Centrifugation

A centrifuge, utilizing centrifugal force to multiply the effect of the difference in specific gravities, is effective in separating solid particles as well as tramp oil from a lubricant. Furthermore, the centrifuge is effective over a wide viscosity range, and with a self-cleaning design, solid particles can be discharged on a continuous basis. Care must be taken in designing the centrifuge cycle so that only tramp oil is removed, not desirable emulsions (Fig. 12.1d).

Low-speed centrifuges are designed specifically to remove suspended solids from any type of liquid, except tramp oil. High-speed units, on the other hand, may be used to remove tramp oil as well as solids. Bacteria and mold, if agglomerated, may also be removed with the tramp oil.

Chip ringers, which are a variation on the standard centrifuge design, separate metal chips from cutting fluids through the use of a rotating basket, first rotating at low speeds, then at high speeds. These alternating cycles separate the fluid from the chips, after which the chips fall into a collecting hopper. Compressed air is used to mechanically clear the basket of sticking chips.

Generally, because of their relatively high capital cost, centrifuges are most cost effective when they are components of a clarification system used on a production line operating at high production rates.

As in the case of hydrocyclones, the use of tank gravity-settling systems, which remove bulk contaminants prior to centrifugation, improves the cost effectiveness of centrifugation.

12.1.5 Magnetic Separators

Magnets may be incorporated into various lubricant clarification systems. These devices are particularly successful in applications in which relatively fine swarf is present, such as that deposited from grinding operations. The part of the swarf that can be magnetizied can be removed by magnetic separators. The nonmagnetic portion must be removed through other procedures, such as filtration.

In one design, a drum that contains a magnetic inner lining attracts the magnetic metallic particles to the surface of the drum as it rotates (Fig. 12.1e). A scraper bar removes the buildup of magnetic materials. The magnetized particles that cling to the drum surface act as filtering media for nonmagnetic particulates. If necessary, a magnetic separating device and a filtration unit may be used in tandem to remove both magnetic and nonmagnetic particulates.

The magnetic separator generally removes particles with a size that exceeds approximately 30 μm, although there have been some reports of successful removal of particles as small as 5 μm. Furthermore, these devices are generally most successfully used when fluid viscosity is relatively low, for example, of the same order as the viscosity associated with emulsions or water-based solutions.

Conveyors that are magnetized have also been used successfully to remove chips from the fluid in operations involving metal removal. A portion of the conveyor is equipped with submerged sealed magnets that carry the magnetic materials to a drop-off discharge portion. In the meanwhile, fluid falls through the conveyor to be recycled.

Variations and combinations of these and other mechanical devices may be used when magnetic separation is appropriate for clarifying the metalworking fluid.

12.1.6 Filtration

In positive filtration the metalworking fluid passes through a filter medium driven by gravity, pressure, or vacuum. Solid particles larger than the openings in the filter medium deposit on the filter, forming a filter cake. The most common filter media used in the metalworking industry are composed of cloth, paper, organic polymers, or wire screens.

Positive filters are most frequently used and operate on the same principle as the gravel and sand filter media used for clarifying water. The large swarf collects on the filter media and forms a filter cake, which in turn filters the smaller swarf and grit from the fluid. The thicker the cake, the finer is the filtration. The filter medium acts as a support for the filter cake.

The efficiency of the unit increases as the cake builds up. Eventually, the density of the cake reaches the point at which the resistance to flow exceeds the vacuum, or pressure, required to maintain the flow. At a predetermined point, the pressure or vacuum is stopped or a level is reached at which the filter medium indexes, exposing fresh medium, and the cycle is repeated. There are various filter systems. Some of the more frequently used systems are described below.

Gravity Filters

Because of gravity, the weight of the metalworking fluid is sufficient to provide the force required to penetrate the filter medium. The fluid cycles from the machine reservoir to the filter unit, where it passes through the disposable filter medium, supported on a steel belt or wire screen, leaving the swarf on the medium. After filtration the clean fluid is recycled for further use. As the cake builds up, the resistance to flow increases and causes the liquid level in the reservoir to rise. A float control activates the belt or conveyor holding the filter medium so that it indexes, exposing a new filter surface.

The advantages of such a system are principally those of low capital cost and ease of operation. The disadvantages are associated with sizable floor space requirements, the need to dispose of relatively more media and accumulated cake, and the necessity to use a metalworking fluid that has low foaming characteristics to reduce filter medium usage.

Pressure Filters

A typical pressure filter contains two horizontal compartments: a movable top compartment and a stationary bottom compartment. During operation, air pressure seals the two compartments together. The filter medium may be a continuous nylon belt used as is or one coated with a disposable medium.

Typically, the system consists of a filter unit, a settling tank or sump, and a drag-out chain. The metalworking fluid is pumped to the top shell of the filter under pressure, forcing the fluid through the filter medium. The particulates deposit on the filter medium to form a cake, permitting the cleaned fluid to accumulate in the bottom shell. The fluid is then pumped to a clean supply tank for reuse.

At a predetermined setting, usually 6 to 9 psi (40 to 60 kPa), the index cycle begins. The pump that supplies the metalworking fluid stops, and compressed air blows the fluid from the top shell through the filter medium to the bottom shell. At a predetermined pressure, the upper shell rises; the filter medium advances, exposing a new surface. The shell then closes, and filtration continues. With disposable media, the accumulated cake is fed to a waste tank. If a continuous nylon belt is used alone, compressed air may be used to blow the swarf into the waste tank.

The principal advantages of such a system consist of its ability to remove relatively small particles efficiently. Furthermore, large fluid volumes can be handled with minimal floor space requirements. On the other hand, tramp oil, if not removed, may prematurely plug the filter medium. Hard-water soaps, if formed, may also plug the medium. For efficient operation, chips as well as the smaller fines should be agitated and pumped to the filter unit. Disposal media usage is high because it is advanced along the entire length of the shell. Maintenance costs may also be quite high.

Vacuum Filters

The most common positive filter is driven by a vacuum and is often referred to as a suction filter. A typical system may consist of two tanks with a filtering chamber at the bottom of one, covered with either disposable or permanent filter media.

The unit with the disposable medium is designed so that the lubricating fluid is piped to the filter area, where it passes through the medium, leaving the swarf behind. As the cake thickens, resistance to flow increases and causes the vacuum to increase. At a predetermined vacuum level, the vacuum switch activates a conveyor, which moves the dirty medium toward the discharge end, exposing a clean section. During this index cycle, the filtered fluid is piped from the clean tank for further use.

The principal advantages of this system include relatively low capital cost and efficient filtration of particulates from 10 to 25 μm in size. The disadvantages are related to the need for relatively large floor space, the cost of medium disposal, and the need to minimize formation of hard-water soaps that can plug the filter medium.

The positive filter with a permanent medium operates on the same principle. The fluid is pulled through the medium by a vacuum, and a cake is formed on a wire screen. A scraper bar, attached to a chain drag-out, stretches across the wire screen medium. When the resistance to flow reaches a predetermined level, the vacuum is turned off, and the conveyor bar is activated and moves about 1.5 to 2 in. (38 to 50 mm), scraping the swarf off the screen in that area. The vacuum is again activated, and the cycle is completed. The swarf is carried away on the drag bars.

The principal advantages are low capital costs and efficient filtration of the particulates. Furthermore, there is no disposable medium cost. The disadvantages are the relatively large space requirements, that fluid selection may be critical to avoid medium plugging, and that the system is not suitable for use in grinding cast iron.

Tube and Leaf Filters

These filters may operate with the application of pressure or vacuum. A compartment contains the filter tubes or leaves, usually made of tubular nylon or woven wire. The metalworking fluid is pumped to the compartment through the filter elements. The clean fluid is then returned to a reservoir for further use. As a cake forms on the filter elements, the flow is restricted, causing a rise in pressure or liquid level, depending on the design of the unit. At a predetermined pressure setting or liquid level, the filter automatically goes through a blow-down or back-flushing index cycle, when compressed air or the clean fluid lifts the cake off the outside of the elements and drops it onto a conveyor for disposal.

When very clean fluid is needed or when particulates are very small (<1.0 μm), the filter may be coated with diatomaceous earth to improve filtration.

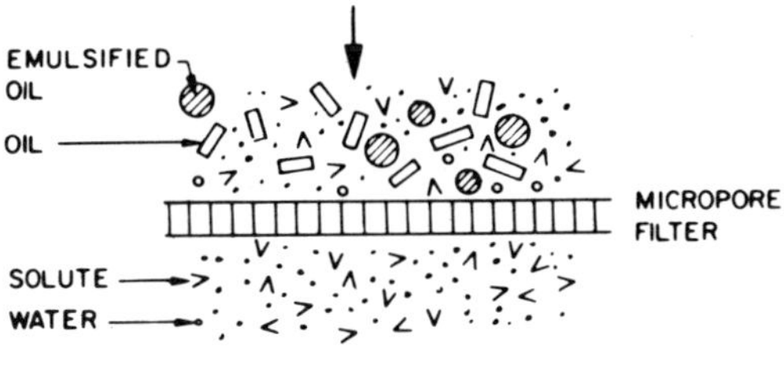

Figure 12.2 Schematic illustration of ultrafiltration (After J. A. Schey, 1983. *Tribology in Metalworking,* American Society for Metals, Metals Park, Ohio, p. 184.)

The filter must be recoated after each back-flushing cycle. Diatomaceous earth improves filtration and in the process not only removes particulates from the fluid but may also remove lubricant components. For this reason, precoating should be used only with metalworking fluids formulated so that they are not susceptible to component removal.

The advantages of such a filtration system are that they are very efficient and can be used when fine finishes are required. The disadvantages are associated with high capital costs, the need to match the metalworking fluid to the system so that the chemistry of the fluid is not deleteriously affected or the filters plugged, and the need to adopt operating practices to minimize plugging of the system.

Ultrafiltration and Reverse Osmosis

Ultrafiltration is used to remove oily components from an emulsion by passing the lubricant through a membrane that retains oil, suspended solids, and bacteria while water and other small molecules pass through (Fig. 12.2). The ultrafiltration membrane cell size varies from 0.01 to 0.0025 μm. Low pressure is applied to force water through the membrane, a process similar to reverse osmosis. The ultrafiltration membrane acts as a molecular filter, the reverse osmosis membrane separates ions.

Generally, ultrafiltration treatment is preceded by the use of a tank in which oil and solids are removed prior to the ultrafiltration. Belt or rope skimmers can remove the oil, and centrifuges are often used to remove both oil and solids. The waste water is then treated by ultrafiltration. The treated water is generally of acceptable quality for direct sewer discharge.

For ultrafiltration to be successful as a purification method applied to chemical solutions, the composition of the solution must be matched to the

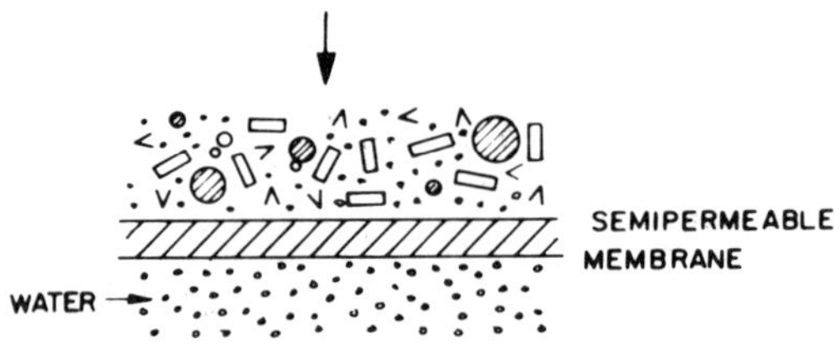

Figure 12.3 Schematic illustration of reverse osmosis. (After J. A. Schey, 1983. *Tribology in Metalworking,* American Society for Metals, Metals Park, Ohio, p. 136.)

filter so that molecular size is compatible. This can be done by appropriate formulation. The reverse osmosis and ultrafiltration processes are similar in that both employ porous membranes. However, the porous membrane pore size in ultrafiltration is generally from 25 to 100 Å in diameter, which passes most inorganic solutes, but intercepts organic solutes of higher molecular weight and suspended solids. Reverse osmosis membranes are composed of materials with pore size of approximately 5 to 25 Å and effectively separate ionic contaminants, such as sodium chloride, from aqueous solutions under pressure-induced osmosis (Fig. 12.3).

Reverse osmosis systems require pressures on the order of 300 to 400 psi (2 to 3 MPa), as compared with 50 to 150 psi (0.3 to 1 MPa) for ultrafiltration systems. Approximately 95% separation can be achieved. If activated carbon is used as an auxiliary filter medium, then essentially 99% of all contaminant ions may be removed. Disadvantages are operating costs higher than those incurred for ultrafiltration, operation of a high-pressure system, and the potential requirement for auxiliary filtration. Advantages are that the water permeate is "pure." Waste oil can be disposed of that is free of contaminating water.

Coagulation

Lubricants used in the processing of aluminum and aluminum alloys build up fines and metallic soaps over time. A method for coagulating the soaps and fines promotes their separation from the metalworking fluid by centrifugation. Coagulation is achieved by heating the lubricant or other metalworking fluid in contact with very small quantities of concentrated sodium carbonate solution. The sodium carbonate in contact with the metal fines neutralizes the charge so that coagulation can take place and the suspension is disrupted.

Disadvantages are the introduction of a chemical process into a metalworking environment. Advantages are low capital costs, with most of the cost associated with purchase of a decanting centrifuge, major reductions in waste volume over commonly used filtration methods, and operating costs much lower than those associated with standard filtration procedures.

Electrostatic Separation

A bed of glass beads surrounding a high-voltage electrode is enclosed in the container through which the metalworking fluid flows. The method has been used successfully on lubricants used in rolling of aluminum; particulate matter of 0.05 to 0.5 μm can be removed. By utilizing an appropriate back-flushing procedure, continuous operation can be realized. The back-flushing fluid is centrifuged, and the resulting semidry cake can then be disposed of as a solid waste.

Disadvantages are relatively unfamiliar technology and relatively high capital costs, including those associated with the centrifuge. Advantages are recycling of particulate-free metalworking fluid and reduction in oil disposal volume, improvement in mill productivity in rolling aluminum, and possible improvement in foil quality.

12.2 RECYCLING

Recycling metalworking lubricants is, of course, only pertinent to those processes in which a continuous flow of lubricant circulates through the system. In Chap. 6 the chemistry of lubricants is described, some of which may be recycled. Obviously, for recycling to be effective, the chemistry as well as physical properties of the fluid must be maintained. Appropriate controls are discussed in Chap. 10.

Assuming that control of properties over appropriate manufacturing time cycles is achieved, and that particulates and other matter present in the fluid are removed with appropriate equipment, one must, in addition, consider other aspects of recycling technology. System design, tankside additive practice, water treatment, and processes for controlling microbial growth are described in the following sections.

In designing the plant recirculation system, a number of important design criteria must be observed.

12.2.1 Plant Air

If air surrounding the work site is not controlled, there are several possibilities for introducing operating problems into the manufacturing cycle, which are often erroneously attributed to the lubricant. For example, acid fumes from nearby pickling, plating, or coating facilities may cause rusting on an intermittent basis. High humidity similarly can result in rust. Plant spores, if not removed, may enter the lubricant stream and cause biological growth. Therefore, sensitivity to the possible impact of contaminated air on plant operations can reduce operating costs, particularly if steps are taken to correct the problem.

12.2.2 Mechanical Design

Lubricant reservoirs should be constructed of materials that are not subject to chemical attack. Concrete tanks are particularly poor choices, since often lime in the concrete can be eroded, particulate contamination is increased and lubricant chemistry can be altered.

Flexible piping should be used only if there is no possibility of buildup of microbial contamination on the walls of the pipe that cannot be readily removed by flushing. All pipes should, along with their attendant valves, contain as few bends as possible to facilitate cleaning.

If flumes are to be used, they should be covered so that waste from the plant floor cannot enter. Furthermore, the flume design should be such as to maximize fluid flow, thus eliminating the possibility of solid buildup, microbial growth, fines, or metal chips.

All surfaces in contact with the metalworking lubricant should be as smooth as possible to minimize any possibility of microbial buildup or deposition of metallic or nonmetallic deposits.

The piping and pump systems should be sized so that lines are completely filled with fluid during operation of the system. Otherwise, slime may develop on the top of the inner wall of the pipe. This slime is very difficult to remove and, therefore, may serve as a source of contamination of any newly added lubricant. If the system is large, pipelines should be constructed so that cleaning pigs can be introduced. Also, access to steam lines should be provided for cleaning in the initial design of the piping system.

Equipment design should be such as to eliminate the possibility of stagnate pools of lubrication. If they do not have ready access to the circulation system, such pools provide an ideal medium for microbial growth.

Reclamation of metalworking lubricants should be carried out on an ongoing basis so that dead storage of used lubricants does not occur, again to eliminate excessive microbial growth and deterioration of the fluid.

12.2.3 Additive Practice

It is a relatively common practice to rely on periodic additions of the neat, or concentrated, lubricant to a recirculating system to replenish and maintain the chemical components, to extend the life of the lubricant with its original properties. However, particular components of the metalworking lubricant may be lost (by evaporation, degradation, chemical reaction, and other factors) at different rates. Under such circumstances it may be cost effective to add those chemical components that are lost, without adding the total package comprising the metalworking lubricant. Some of these additives are as follows.

Emulsifiers

Metals to be formed are often coated with oil applied at the hot or cold mill. This coating may often enter and contaminate the lubricating fluid, or leakage of hydraulic system oils may also contaminate the fluid. When the lubricant is an emulsion, emulsification of the contaminating tramp oil and its attendant additives may be desirable. Tankside addition of an emulsifier, with appropriate chemistry, can extend the life of the fluid. The tramp oil then becomes incorporated as a component of the emulsion.

This practice may have associated deleterious operating problems because of the incorporation of foreign additives in the lubricant, which may cause various problems. Rust, staining, deposits, and dermatitis may result. Obviously, various other contaminants, such as grease and solvents, may also enter the system and react with the emulsifier, all potentially having adverse operating effects. Therefore, the institution of such a practice should be carefully monitored before adopting it on any large scale.

Chelating Agents

NaEDTA (disodium ethylenediaminetetraacetate) is used frequently in emulsions because it effectively complexes cations, such as calcium, magnesium, and aluminum, thus improving the stability of emulsions, preventing fatty acid oxidation (when fatty acids are present in the formulation) and the precipitation of insoluble salts. Furthermore, chelating agents may potentiate the properties of some biocides. The generation of foam may result from such tankside additions, and therefore, if it occurs, attention should be paid to the possible use of harder water or antifoaming agents.

Antifoaming Agents

Various chemical compounds may be added to fluid systems, causing, for one reason or another, foaming to take place. Antifoaming additives reduce surface tension and can be very effective when used in small quantities. Silicon antifoaming agents are expensive yet frequently are cost effective. Kerosene or very low viscosity mineral oils also can be effective when their addition does not create operating problems, such as coatings or increased cleaning costs.

Biocides

The biology of metalworking lubricants is discussed in Chap. 11. Tankside additions of biocides can be very effective in prolonging the life of the lubricant. Care should be taken to add the biocide only when required, not in excess. Excess biocide can be harmful by changing the pH of the fluid excessively, as well as through sensitization of the skin of employees in contact with the lubricant. Furthermore, it is important to be aware that if the microbial content of a fluid is high when treated with a biocide, the dead microbe concentration can lead to excess foaming within 24 to 48 hr and so should be removed. Furthermore, the excess dead microbes can load filter systems unduly and may promote undesirable deposits.

Wetting Agents

Because of metal-lubricant interactions, as well as extraneous chemical systems interacting with the lubricant, the wetting characteristics of the lubricant may change over time. An appropriate wetting agent can be added, depending upon the chemistry of the lubricant, to restore desired wetting characteristics.

Water

It is important to emphasize that periodic addition of water to the lubricant system may have very important effects. For example, if the water is hard, that is, contains high levels of dissolved solids, microbial attack on the fluid can be accelerated, corrosion may occur, and deposits may reduce filter life and coat the equipment and workpiece. Appropriate water treatments must be investigated in terms of their effect on the lubricant. There is some evidence that suggests that softened water may foam excessively. Water-treatment procedures are described in detail in Sec. 12.3.

Deodorizers

Under certain operating conditions an unpleasant odor may result, due to degradation or contamination, that should be compensated for by the addition of an appropriate deodorant. The deodorant must be selected to be compatible with the system.

Dyes

If the lubricant is given a distinctive color for control purposes during processing, it is important to maintain the identifying color during extended use. Addition of an appropriate dye may be practiced for this purpose. The effect of a dye on workpiece surface appearance should be evaluated prior to a final processing decision.

Lubricant Components

Since various chemical components of a lubricant are depleted at varying rates, depending upon various interacting phenomena during the metalworking operation, analysis of these phenomena can lead to a practice for replenishing the components based upon need. Fats, polyglycols, chlorinated paraffin, and other substances may have to be added to the lubricant to maintain the desired operating characteristics. Cooperation between the user and lubricant supplier is essential in the analysis of and a decision for changes in the operating practice, if any, that should be instituted.

12.3 WATER TREATMENT

Water-based solutions and emulsions are widely used as metalworking lubricants. Soluble inorganic compounds, such as calcium chloride, aluminum sulfate, and iron sulfate, provide food for microbial growth. Furthermore, they may cause corrosion or staining, may deposit sticky gummy residues, and, finally, can upset the chemical balance and cause splitting of emulsions and the deposition of chemical components from both solutions and emulsions.

Organic contaminants in the water, which may partially consist of microbes, such as bacteria and mold, may, in their growth, lower the pH of the fluid, accelerate corrosion, promote unwanted deposits, and, finally, cause premature plugging of the filtration system.

Various techniques have been developed to treat the water to eliminate, or at least minimize, these effects. The most commonly used processes are described below.

12.3.1 Commonly Used Water Treatments

Water Softening

Water is passed through an ion-exchange resin. The sodium ions of the resin replace the calcium and magnesium ions present in the water, thus resulting in softened water. Periodically, the ion-exchange material is regenerated by back flushing with a saturated sodium chloride solution. The total amount of solids present in softened water is not appreciably affected, but the nature of the water is changed since the hardness has now been significantly reduced. Softened water may have a tendency to foam excessively.

Reverse Osmosis

Processed water is passed through a semipermeable membrane that allows passage of the water molecules but does not permit passage of the soluble salt contaminants. Approximately 90% of the minerals present can be removed from the water by this process. If the initial water is relatively pure, the economics of reverse osmosis may be attractive. On the other hand, membrane life can be unpredictable and replacements are costly.

Deionization

In this process, both cations and anions, which deleteriously affect the water, are removed by separate exchangers. One is a cation exchanger that replaces the normally occurring sodium, potassium, iron, magnesium, and other cations in the water with hydrogen ions, and the other an anion exchanger that replaces the sulfates, chlorides, and carbonates with hydroxyl ions. These exchanger units can be regenerated by appropriate backwashing with hydrochloric acid and sodium hydroxide to regenerate the cation exchanger and the anionic exchanger, respectively. The water quality produced is comparable to the water resulting from a distillation process. The cost is very competitive and varies from about 0.2 to 2.0 cents per gallon.

Distillation

Processed water can be treated by evaporation and condensation of the evaporated water. The salts precipitate, and the water vapor condensed is very high in purity. The process may be carried out with both high pressures and temperatures to accelerate the recovery rate. Boiler water, or rainwater, if available, may serve as an adequate substitute for distilled water.

12.3.2 Control of Microbial Growth

The usefulness as well as some of the potential problems associated with periodic additive treatments have been discussed, and various technologies useful in maintaining the general cleanliness of the recycling fluid have been reviewed.

Various methods for the continuous treatment of fluids to eliminate microbial contamination are briefly described below.

Ultraviolet Radiation

Pretreatment of water to be used in those cases where the metalworking fluid is at least partially composed of water has been successfully carried out using ultraviolet light to prevent microbial contamination. The use of microbial agents is eliminated, and furthermore, no physical or chemical changes are induced in the water. However, for this technology to be effective, the water must be free of particulates. The ultraviolet units are relatively compact and can be installed on line ahead of the point of use. Frequently, a process consisting of deionization, followed by ultraviolet radiation, provides noncontamined water to the makeup stage of the particular metalworking fluid.

Radioactive Irradiation

An alternative to ultraviolet radiation is the use of low-level radioactive irradiation to prevent microbial contamination. The process is essentially the same as that for ultraviolet radiation. The method is effective, although not yet widely used, primarily because of society's concern about atomic processes such as irradiation.

Thermal Treatment

During recirculation the metalworking fluid may be heated to temperatures on the order of 160°F (70°C), thus killing microbes present in the fluid. Furthermore, the heating step facilitates the separation of tramp oil from the metalworking fluid so that it can be more readily removed. Solid contaminants also tend to separate and deposit in the bottom of the holding tank or sump, as the case may be, for purposes of separation and removal. Obviously, the heat source can be selected on the basis of cost effectiveness. Furthermore, the application of heat may be at the water makeup stage or during the recycling of the used fluid, depending upon the particular circumstances.

12.4 DISPOSAL

If the level of contamination in the lubricant exceeds established limits, the lubricant should be disposed of and the lubricating system and equipment should be cleaned and recharged with fresh lubricant. The decision of excessive contamination involves studying the relationship between contamination and undesirable processing effects, such as corrosion, bacterial degradation, part staining, and reduced tool life. A value analysis of the various cost benefits involved in the lubrication life cycle will establish appropriate guidelines for disposal.

12.4.1 System Cleaning

When the disposal decision has been made and the metalworking fluid is removed, the system should be cleaned thoroughly before recharging it with new lubricant. Frequently, machine cleaners are added to the lubricant without due regard to the effect on both productivity and the potential increase in contaminant load as a result of the cleaning of pipes and equipment. In general, a practice should be followed that involves removing the coolant and precipitated particles as completely as possible, followed by circulation of a machine cleaner through all the pipes and valves and all traps and pans. All filters should be cleaned or replaced. Subsequently, steam cleaning, if feasible, or hot water flushing of the system should be carried out, followed by a flush of the system with an effective biocide.

During any one of these stages gross contaminants can be loosened, particularly mold or sludge, and the cleaning practice should be continued and extended until such materials no longer appear in the cleaning solution.

Interactions between retained cleaners or biocidal flushes with new metalworking fluids should not be permitted. They can be eliminated either by choosing cleaners and biocides that are compatible with the metalworking fluid and/or thorough and complete flushing with hot water or steam prior to recharging the system.

12.4.2 Splitting Emulsions

If the metalworking fluid is an emulsion, disposal costs will be reduced if the water phase and any solid particulate phase can be separated from the organic or inorganic components of the lubricant. Splitting the emulsion may be achieved in many ways. Electrolytic treatment of the fluid, thus breaking down the attractive forces between the water and oil phases, can be effective. Heat may be applied and can, in some cases, effectively split the emulsion.

Generally, however, the practice used most commonly consists of chemically treating the emulsion to nullify the attractive forces, destabilizing the emulsion, and thus causing splitting. A practical cycle for such treatment might consist of the following steps.

1. Separate tramp oil and solid particulates, using methods previously described.
2. Using appropriate safety equipment, cautiously add concentrated sulfuric acid to the emulsion until the pH is approximately 3.5. The solution should be mixed constantly while the addition is being made.
3. Add aluminum sulfate to the acidified emulsion while mixing. The concentration of the aluminum sulfate to be used depends upon the volume, chemistry, and pH. However, it must be sufficient that precipitated aluminum hydroxide forms.
4. Recycle the emulsion back to a pH of between 6.5 and 7 by adding 50% sodium hydroxide to the emulsion, again with care and using safety equipment.
5. Allow the mixture to stand to permit aluminum hydroxide floc to settle.
6. The clear water layer should now be pure enough for disposal. The aluminum hydroxide floc can be removed and treated with concentrated

sulfuric acid to regenerate aluminum sulfate, which, with new aluminum sulfate, can be used for the next splitting cycle.

12.4.3 Fluid Disposal

Portable sump cleaners are frequently used to physically remove spent metalworking fluids. These portable cleaners remove lubricants, chips, and swarf from the sumps. Depending upon the design, the filtration, settling, and pretreatment of the sludge may take place in the unit before it is moved to a central collection facility.

Oils are split from the metalworking lubricant, generally at the central collection site, as previously described. The water fraction may be pretreated prior to recycling or disposal. Chips and sludge are treated separately, as described in Sec. 12.4.4. The oil may be suitable for collection by the re-refiner, or it may be used internally for less critical lubrication applications or as a fuel.

12.4.4 Sludge and Chip Disposal

The sludge containing waste products must be disposed of in an environmentally acceptable manner in government-approved sites. The impact of government regulations on sludge disposal is discussed in Chap. 15.

Chips produced by a machining operation, on the other hand, are often sold back to scrap dealers after removal of as much oil as possible through separation procedures, such as heating and/or the use of centrifuges.

12.4.5 Re-refining

A variety of re-refining cycles has been developed to produce oil equivalent in quality to that purchased from base oil refineries. These processes have been successful in treating waste liquid lubricants used in metalworking. The recovery of oil from emulsions complicates the process. In the past, the re-refining cycles, either because of poor controls or inadequate technology, often produced oils that were inferior in quality. The uncertain supply and rising costs have led, in the past several years, to the development of much improved processes and technology with acceptable control.

Either alone or in combination, hydrotreating and vacuum distillation, with or without pressure, have been used to produce desirable base oil stocks from used oil and eliminate the formation of toxic solid wastes. In general, toxic wastes are related to the necessity for removing aromatic organic materials present in the acid clay, which is frequently used to re-refine the waste oil. These processes, which use conventional acid and clay treating techniques, have the associated problem of disposal of the solid waste.

The economics of re-refined base oil is as yet unclear. However, it appears that, in the long term, re-refining will grow in importance.

BIBLIOGRAPHY

Bennett, E. O., 1976. *System Design and Housekeeping Practices as Factors in Prolonging Coolant Life,* Department of Biology, University of Houston, Texas.

Fritsche, G. R., and S. J. Sanosonetti, 1980. Electrostatic separation of aluminum fines from aluminum foil roll-oil lubricants, *Lubrication Engineering,* 36(3), 154-159.

Green, B., 1982. *Ultrafiltration: Capabilities and Limitations,* SME Technical Paper MS82-921.

Hoefling, E., and J. Oertle, 1981. Coagulation – a new method for the cleaning of aluminum cold rolling mills, *Lubrication Engineering,* 37(4), 195-197.

Meares, P., 1976. *Membrane Separation Processes,* Elsevier Scientific, New York, pp. 81-84.

Swain, Jr., J. R., 1983. Concentration, recycling and disposal of industrial lubricating fluids, *Lubrication Engineering,* 39(9), 551-554.

Wick, C., 1977. Clean coolants cut cost, *Manufacturing Engineering,* November, 34-40.

13
Delivery, Storage, and Distribution

Planning and organizing cost-effective delivery of lubricants to a plant and, subsequently, within the plant to the point of use is an important aspect of metalworking operations. Contamination, deterioration, control, and handling requirements are all responsive to delivery and distribution organization to and within the plant. Storage of lubricants requires planning and organizing to control use, protect quality, and provide inventory control.

The direct cost of metalworking lubricants may be reduced through volume purchases of multiple drum quantities or through bulk purchases. However, these direct cost savings can be negated through hidden handling costs, increased inventory, or a reduction in quality as a result of contamination and/or deterioration during storage.

13.1 DELIVERY

Although information on the relationship between volume purchasing and cost is readily available from the supplier, delivery costs to the plant receiving dock are not as easily obtained. Truck and/or railcar delivery costs vary, depending upon the carrier and quantity to be shipped. The necessity for heated carriers in winter to eliminate potential splitting of water-based concentrates, or precipitation of additives from oil- or solvent-based lubricants due to freezing, needs to be evaluated. Economical order size, as well as assurance of on-time delivery and supplier control of quality in the delivered package, are all subjects of importance relative to the purchase decision.

Lubricant stock in a plant should be adequate but not excessive. Ease of delivery in summer and difficulty in winter should be taken into account. There should be an agreed minimum stock for each type of lubricant stored, and replacement should be provided for in advance.

Receiving lubricants should involve acceptance or rejection of the container as well as its contents. How extensively this inspection should be carried out depends upon the history of previous rejection rates as well as direct inspection costs. Leaking drums can result from using defective drums during initial filling, as well as from damage caused by subsequent handling. Handling, cleanup, paperwork, and loss of product all increase costs.

In planning a lubricant storage and distribution center, the method of delivery is an important consideration. Usually, lubricants are delivered by truck. Thus, facilities should be made available for safe unloading of the heaviest package, usually a full drum of 55-gal (208-liter) capacity. If it is unloaded without proper equipment, there is danger to the personnel, to the package, and to its contents.

Some trucks are fitted with tailgate hoists so that drums can be off-loaded at floor height. A suitable drum-handling device, such as a drum skid, should be made available to the personnel handling the drum. In the unloading area, an overhead hoist running into the central storage area is of considerable assistance in both unloading and stacking. A forklift truck is most useful and can speed safe transport of the relatively heavy drum.

If there is an unloading dock at the average height of the delivery truck platform, a drum hoist may be used. After being unloaded, the entire consignment should be checked against the invoice. The condition of the drum should be observed, and any leakage should be reported.

When oil is delivered in bulk, the unloading operation requires little labor. The tank truck or railcar volume or weight is measured and compared to the shipping invoice. If the invoice is correct, the truck or railcar is unloaded through an appropriate hose and pump hookup.

Although a reduction in bulk prices, when compared with drum prices, is attractive, as is a reduction in handling associated with bulk storage, a price is exacted that should be evaluated as part of the value analysis process. If the volume of usage justifies installation of bulk tanks and receiving as well as in-plant delivery systems, then a value analysis of all aspects of such a decision should be carried out. Such advantages as bulk pricing and reduction in handling costs associated with drum usage (loading and unloading and in-plant movement) should be considered, as well as the reduction in storage space and waste, elimination of drum deposits, and the cost of lost or damaged drums.

As is generally the case in manufacturing decisions, however, costs as well as benefits are associated with bulk storage of lubricants. A capital cost for tanks, gauges, piping, valves, and pumps is required. Space must also be made available. The entire installation must be in compliance with applicable federal, state, and local safety, fire, and contamination regulations. Any insurance implications must also be evaluated with respect to both coverage and cost.

Errors in product quality or composition, contamination of the product, mistakes in delivery by the supplier, or to the wrong bulk tank internally, are all potential problems. These can occur during operations and are magnified by the volumes involved. Furthermore, unless there is an appropriate distribution system for bulk lubricants within the plant, distribution of drums to work sites may be a more effective method of internal control and distribution.

13.2 STORAGE

During storage, lubricants must be protected from contamination and deterioration as well as from physical damage. Collection of water inside a sealed drum or storage tank is a frequently misunderstood problem, leading to deterioration of its contents. All storage vessels (drums or tanks) breathe with changes in temperature. Water may enter the container because it collects on top of the drum, or through condensation of water from the air external to or within the drum. Changes in relative humidity and temperature, particularly abrupt changes, can result in water precipitation within the confines of both the drum and the tank. Over a long period of time, airborne moisture can enter a container even when the container appears to be firmly sealed.

Elimination of water contamination of sealed drums is important in maintaining content quality as well as drum integrity. Water vapor inside a sealed drum can cause rust, with subsequent lubricant contamination and acceleration of the deterioriation of the lubricant. Chemical reaction within the drum caused by water can be best avoided by storage of the drums in an area where temperature and moisture changes are slight and freezing temperatures are avoided. If the drums are stored where water may collect on the tops and temperature variations are unavoidable, then the drums should be stored tilted so that the contents seal the inside of the bungs, creating a barrier to the entry of water or water vapor. Tank storage involves similar water problems, hence the importance of appropriate venting and periodic cleaning.

All too often, when drums are stacked, the top drums are used and replaced by new drums, whereas the lower drums remain unused for long periods of time. Depending upon the chemical as well as physical characteristics of the lubricant, the viscosity may change with time and some of the constituents may precipitate. For this reason alone, racking is to be preferred to stacking to ensure first-in, first-out movement of the drum. Well-designed racks save space, ease handling, and ensure first-in, first-out procedures.

Extremes of temperature may deleteriously affect the performance of metalworking lubricants. If temperatures are low, particularly below freezing, the contents of the containers may, at the very least, become unduly viscous and thus difficult to distribute. In the case of oil- or solvent-based lubricants containing fatty additives or solid lubricants, these additives may precipitate. High temperatures for sufficiently long times may result in deleterious oxidation of the oil content of the lubricant, unduly evaporate the water in an emulsion, oxidize the product, and result in deterioration of the additives. Generally, temperatures between 40 and 100°F (4 to 38°C) are safe.

Central storage of lubricants purchased in drum quantities is desirable, from the viewpoint of controlling their use and environment as well as protection of the container and elimination of potential contamination. It is important that adequate means for distributing the lubricant from the drum also be made avilable. Handling equipment must be simple and compact. Racks or shelves should be provided for measuring and other dispensing equipment. Containers and measuring equipment should be kept clean; hence, a means of cleaning them should be provided.

Pumps are sometimes used for dispensing lubricants from drums, and in this application, the drum stands on end with the pump fitted in the bunghole. To avoid contamination, pumps, if used for pumping more than one product, should be cleaned after each use prior to subsequent use.

There should be a drip tray below each drum tap, and any drips or spillage on the floor should be cleaned without delay. Measuring cans with easy-pour spouts are needed so that the amounts of lubricants issued can be recorded. If drums are to be distributed to work stations from which concentrate is to be dispensed to a reservoir, appropriate means must be taken to safeguard both the integrity and the cleanliness of the concentrate as well as the distribution system to the reservoir.

If bulk tank storage is to be made available, the appropriate capacity should be decided upon. Tank design should be proper to provide for both filling and dispersal. Tanks for most lubricants are made of mild-steel plate or sheet. Joints should be welded with reinforced-angle iron frames with an inspection hole with an air vent and filter at the top to prevent ingress of solid contaminants or liquids, such as water, should be provided along with an access manhole for maintenance.

Foundations should be sufficiently strong to carry the weight of the tank and its contents. The tank should be slanted (2 to 3°) toward the pipe outlet. A cutoff valve should be provided at the tank to permit cutoff there as well as in the pipeline. An adequate measuring system for indicating the volume of liquid in the tank should be easily visible and not subject to errors caused by filling or extremes of temperature.

13.3 DISTRIBUTION

Frequently, the care that has gone into protection of both the container and the contents is completely overlooked when lubricants are dispensed within the plant. Pumps, reservoirs, and applicators are frequently overlooked with respect to the possibility of their contamination. Furthermore, plant operations are such that lines are broken and dirt frequently enters the lubricant. (Dirt may consist of everything from old shoes to salt, food, cleaning compounds, mold, bacteria, and metal fines and chips.) The integrity of the fluid must be maintained if adequate performance is to be controlled.

Control of lubricants requires sampling at the point of delivery, the point of distribution, or subsequently during the course of use. During sampling, an overriding rule is not to introduce contamination into either the sampling procedure or the container of the sample. Furthermore, the sample must be representative of the material being sampled. Often, samples are taken at locations or with procedures that do not adequately reflect the true condition of the lubricant.

Samples should be labeled and recorded, as to location, date, and type of product, prior to submission for evaluation. If samples are to be sent by mail, appropriate containers must be used to prevent breakage and ensure safe arrival.

Control and distribution of the metalworking lubricant is a prerequisite to its economical application and use. The savings inherent in bulk purchasing can be completely lost if the costs of control and internal delivery are excessive. On the other hand, the convenience of 55-gal drums, located where they are used, provides an efficiency of application that can override the bulk-purchasing cost savings. In this latter case, however, control and appropriate protection of the quality and integrity of the lubricant and its container must be provided.

The return of used containers deserves attention to maximize the percentage of acceptable drums that can be rebated by the supplier. Drum integrity should be maintained. Drums should be emptied prior to sealing and should be stored where they can be picked up in some reasonably controlled manner.

Recent laws and regulations governing disposal of waste oils require adequate attention to disposal. It is not appropriate, within the meaning of the law, to return waste oil to the lubricant supplier unless the supplier has been certified as a waste oil scavenger.

Used drums should never be utilized as containers for different materials without thorough prior cleaning. Chemical reactions inside the drum can lead to explosion of the contents. If drums have had waste oil stored in them and are subsequently sealed, they may become dangerously explosive if temperatures rise and the contents become heated.

BIBLIOGRAPHY

"Do You Know" Bulletin, 1981. Tower Oil & Technology Company, Chicago, Illinois, September.

Huth, W. J., 1974. Storing lubricants, *Lubrication Engineering,* 30(3), 119-127.

Minetola, A. J., 1981. Bulk lubricant storage and distribution, some of the problems, *Lubrication Engineering,* 37(5), 268-271.

Wills, J. G., 1980. *Lubrication Fundamentals,* Marcel Dekker, New York, Chapter 15.

14

Cleaning and Removal of Lubricant Deposits

Metalworking lubricants may need to be removed from workpiece surfaces prior to secondary operations, such as painting, welding, or plating. The parts produced may be covered with soils that have an extremely complicated composition. Slushing and quenching oils, cutting fluids, drawing compounds, inorganic salts, rust, and other oxidation products, for example, may comprise the soil. Oil, grease, and soap, as well as metal chips, may also contaminate the surface of the part. Various cleaning methods that have been developed for removing these undesirable deposits are described in this chapter. This chapter concerns only the cleaning that is relevant to lubricant removal. This chapter does not deal with the important aspects of oxide and scale removal from parts or the removal of soils other than those related to the lubricant film.

The development of lubricant solutions (synthetic) that do not contain oils, as well as lubricants using evaporating solvents, have led to a significant reduction in the need for cleaning prior to secondary operations (Table 14.1). The lubricant film itself may not interfere with welding or painting subsequent to part production. However, the use of lubricants, compounded or otherwise, often results in a deposit that must be removed prior to subsequent operations.

Frequently, cleaning installations are designed primarily to remove oxide scale. Little attention is paid to the removal of deposits, which may vary as lubricants and/or other surface soils change. The specific requirements of cost-effective cleaning of parts with surfaces covered with lubricant frequently receive little or no attention prior to occurrence of a plant problem.

Table 14.1 Metal-Cleaning Processes for Various Purposes (Processes Listed in Order of Decreasing Preference)

Type of production	In-process cleaning	Preparation for painting	Preparation for phosphating	Preparation for plating
		Removal of pigmented drawing compounds[a]		
Occasional or intermittent	(1) Hot emulsion hand slush, spray emulsion in single stage; or (2) vapor slush degrease[b]	(1) Boiling alkaline, blow off, hand wipe; or (2) vapor slush degrease, clean[c] hand wipe; or (3) acid	Hot emulsion hand slush, spray emulsion in single stage, hot rinse, hand wipe	Hot alkaline soak, hot rinse (hand wipe if possible), lectrolytic alkaline, cold water rinse
Continuous high production	Conveyorized spray emulsion washer	Alkaline soak, hot rinse, alkaline spray, hot rinse	Alkaline soak, hot rinse, alkaline spray, hot rinse	Hot emulsion or alkaline soak, hot rinse, electrolytic alkaline, hot rinse
		Removal of unpigmented oil and grease		
Occasional or intermittent	(1) Emulsion dip or spray; (2) vapor degrease; (3) cold solvent dip; or (4) alkaline dip, rinse, dry (or dip in rust preventative)	Vapor degrease or phosphoric acid clean	(1) Emulsion dip or spray, rinse; or (2) vapor degrease	Emulsion soak, barrel rinse, electrolytic alkaline, rinse, hydrochloric acid dip, rinse
Continuous high production	(1) Automatic vapor degrease; or (2) emulsion, tumble, spray, rinse, dry	Automatic vapor degrease	(1) Emulsion power spray, rinse; (2) vapor degrease; or (3) acid clean[c]	Automatic vapor degrease, electrolytic alkaline, rinse, hydrochloric acid dip, rinse[d]

		Removal of chips and cutting fluids		
Occasional or intermittent	(1) Alkaline dip and emulsion surfactant; (2) Stoddard solvent or trichlorethylene; or (3) steam	(1) Alkaline dip and emulsion surfactant; or (2) solvent or vapor	(1) Alkaline dip and emulsion surfactant[e]; or (2) solvent or vapor	Alkaline dip, rinse, electrolytic alkaline,[f] rinse, acid dip, rinse[g]
Continuous high production	Alkaline (dip or spray) and emulsion surfactant	Alkaline (dip or spray) and emulsion surfactant[c]	Alkaline (dip or spray) and emulsion surfactant	Alkaline soak, rinse, electrolytic alkaline,[f] rinse, acid dip and rinse[g]

[a]For complete removal of pigment, parts should be cleaned immediately after the forming operation, and all rinses should be of spray type, where practical.
[b]Used only when pigment residue can be tolerated in subsequent operations.
[c]Phosphoric acid cleaner-coaters are often sprayed onto parts to clean the surface and leave a thin phosphate coating.
[d]Some plating processes may require additional cleaning dips.
[e]Neutral emulsion or solvent should be used prior to manganese phosphating.
[f]Reverse-current cleaning may be necessary to remove chips from parts with deep recesses.
[g]For cyanide plating, acid dip and water rinse are followed by alkaline and water rinses.
Source: *Metals Handbook*, 9th ed., vol. 5, *Heat Treating, Cleaning and Finishing*, 1982. American Society for Metals, Metals Park, Ohio, p. 5.

Fortunately, in most circumstances, minor changes in either cleaning procedures or lubricant composition can promote cost-effective cleaning.

In general, the selection of a cleaning procedure is influenced by the type of soil, the required degree of cleanliness, and the cost of removal. Handling of the part prior, during, and subsequent to cleaning, as well as maintenance of the cleaning solutions and their disposal, are considerations in the overall cost-benefit analysis.

In this chapter we will deal with lubricants, oils, oil-based emulsions, and water-based solutions, as well as coatings, as related to the following five principal cleaning methods that are more or less effective in removing lubricant deposits:

Solvent cleaning
Vapor phase degreasing
Alkaline cleaning
Emulsion cleaning
Acid cleaning

14.1 THE CLEANING MECHANISM

Cleaning consists of one or more of the following processes.

Solution. When the surface deposits are water soluble or solvent soluble (depending upon the cleaning process used), the soil will dissolve in the cleaning solution. Many lubricant compositions are all, or in part, soluble. Formation of appropriate surfactant micelles also assist in this process.

Saponification. This is a chemical reaction that converts animal or vegetable oils to a soap that is soluble in water. Soap formation is a principal mechanism utilized in alkaline cleaning.

Emulsification. This is a process that depends upon the presence of suitable surfactants in the cleaning solution. These react with the lubricant deposits, reduce the surface tension between the cleaning solution and the oily soil, and form an emulsion. The soil and emulsifier then become suspended as a discontinuous phase in the emulsion, when the continuous phase is either a solvent or the water component of the cleaner.

Dispersion. Dispersion of the soil is promoted by specific surface-active materials as components of the cleaning solution. As a result, the concentration of soil on the surface of the part is decreased.

Aggregation. By the addition of appropriate surfactants to the cleaner, lubricants are removed from the surface and collect as large dirt aggregates, which can then be more readily removed. This process of removal is an important part of the overall cleaning process.

14.2 ALKALINE CLEANING

Alkaline cleaning solutions are the most widely used for removing metalworking lubricants, particularly if the subsequent operation is plating.

14.2.1 Composition

Alkaline cleaners are a complex combination of chemicals that are soluble in water. Water, which is the most important component, should not be excessively hard since the presence of hard-water elements reacting with the chemicals in the cleaner reduce their effective concentration and reduce the cost effectiveness of the alkaline cleaning compounds.

The principal chemical components of alkaline cleaners are "builders," several of which are normally used to provide sufficient reserve alkalinity, buffering, and water softening. The principal builders and their characteristics are as follows.

1. Carbonates are low in cost and effective as a buffer and as water softeners.
2. The phosphates, trisodium phosphate, tetrasodium pyrophosphate, and sodium tripolyphosphate, are effective water softeners. Tetrasodium pyrophosphate is a good emulsifier and detergent, as well as a dispersing and deflocculating agent. Sodium tripolyphosphate is the best water softener of the three phosphates.
3. The silicates, sodium orthosilicate, sodium metasilicate, and sodium sesquisilicate, are excellent emulsifiers. They also provide reserve alkalinity and serve to suspend soil after it has been removed from the metal surface. The silicates provide active alkalinity and are good buffers at elevated pH (9 or above).
4. The hydroxide, caustic soda, because of its low cost, is used to supply a significant proportion of the alkalinity required in alkaline cleaning solutions. Furthermore, it increases electrical conductivity and improves saponification.
5. The surfactants, sodium stearate and sodium resinate, are commonly used soaps, although the latter, because of higher cost, is less frequently used. A combination of stearate soaps and synthetic detergent-based soaps is used in most cleaners. Alkyl aryl sodium sulfonate is a widely used synthetic anionic detergent. Sulfonated esters and ethers and polyoxyethylene nonionics are also widely used. Appropriate combinations of anionic and nonionic synthetic detergents are frequently used in the formation of alkaline cleaners.

14.2.2 Equipment and Methods of Cleaning

Different procedures have been developed to utilize the alkaline cleaning process. Some of these methods are described briefly.

Frequently, small parts may be cleaned in rotating drums or barrels. Continuous processing may be carried out by utilizing soak tanks, one for the cleaner and one or two for rinsing with water. Spray cleaning is almost always conveyorized and consists of a steam-heated alkaline cleaner tank, a hot water rinse tank, and a hot air drying zone. Spray nozzles are mounted above and below the conveyor belt to ensure adequate access of the part to the cleaning solution and, subsequently, of the rinse.

Electrolytic cleaning is frequently used as part of an automatic plating cycle. The introduction of electrical current into the cleaning cycle facilitates and accelerates the effectiveness of the cleaner but is attended by high costs and increased electrical hazards.

Spray rotary cleaning consists essentially of a washing machine, such as that used in the home laundry, with cycles of alkaline cleaner and rinse water pumped through the machine, followed by hot air drying. This method is particularly advantageous for processing small lots in a continuous line. Spray rotary brush cleaning is essentially a modification of the spray cleaning procedure, utilizing rotating brushes to introduce mechanical cleaning into the cleaning cycle.

These methods are designed to improve the cost effectiveness of the cleaning process and must be matched to the production requirements as well as the part transfer system available.

14.2.3 Low-Temperature Cleaning

Because of the increased cost of energy, as well as more stringent operating and disposal requirements, attention has been devoted to the formulation of alkaline cleaners that may be used effectively at room temperature. These cleaners are based principally on silicates and phosphates and frequently contain borax, sodium tetraborate, and/or sodium nitrite to inhibit rust.

The normal operating temperatures are on the order of 100°F (38°C) at a pH of from 8.5 to 9.5, as compared with 11.5 to 12.5 for the hot alkaline cleaners. Spray application of the cleaners at pressures ranging from 20 to 60 psi (140 to 410 kPa) improves the efficiency of the cold cleaners.

Parts are clean and cool, and no heating is required. Furthermore, maintenance is reduced. On the other hand, parts become wet and may require drying after cleaning. The cleaning solutions themselves are more difficult to make up and may be more costly when compared with hot alkaline cleaning baths. The cost of heating is reduced.

14.3 EMULSION CLEANING

Generally, emulsion cleaners are used to remove lubricants or other organic coatings from parts prior to subsequent processing. Emulsion cleaners may serve as a precleaning step prior to alkaline cleaning. These cleaners often consist of deodorized kerosene and mineral seal oil in water, emulsified with appropriate emulsifiers. Emulsion cleaners are generally of four distinct types:

1. Stable emulsions, which consist of small particle-size solvent globules suspended in water
2. Unstable emulsions that require agitation to maintain dispersion
3. Multiphase emulsions, which consist of separate phases that are not mixed except during cleaning
4. Emulsifiable solvent systems, which rely upon a water rinse, containing emulsifyers to emulsify a solvent layer previously applied to the lubricant-coated metal surface

The selection of an emulsion system depends upon the type of soil, water hardness, and/or cleaning sequence. Generally, stable emulsion cleaners are the most economical and are particularly useful for removing light metal-working oils while providing some in-plant rust protection. Unstable emulsion cleaners, on the other hand, are frequently used when heavier lubricant films are present on the workpiece surface. These emulsions can also be used more effectively with hard water. Multiphase emulsion cleaners are successful when used to remove lubricant coatings that are more difficult to remove. Surface cleanliness is generally better than the cleanliness resulting from use of other emulsion cleaning systems.

14.3.1 Composition

The solvents used are generally petroleum based and may be paraffinic, naphthenic, or aromatic in composition. The solvents have boiling points ranging from 110°F to approximately 500°F (43 to 260°C).

Emulsifiers consist of sodium or amine soaps of polyalcohols, polyethers, or amine salts of alkyl aryl sulfonates. Frequently, a higher molecular weight alcohol or ester may be added to increase the oil solubility of the emulsifier. A common solvent is mineral seal oil. Commonly used soaps are based on rosin short-chain fatty acids, saponified with potassium hydroxide or organic amines. Glycol and glycol ethers are generally used as couplers to improve the stability of the emulsion.

14.3.2 Equipment and Methods

Immersion and spray cleaning methods are generally based on the size and shape of the part to be cleaned. Spray cleaning is generally more suitable for very large parts that are not intricate in shape. Immersion cleaning is preferred for small parts that can be placed in baskets. Tank agitation assists in soil removal. Combinations of spray cleaning and immersion are also often used. Generally, the immersion tank is followed by a rinse tank.

Spray cleaning units are designed to have separate zones for cleaning, rinsing, and forced air drying. The efficiency of a spray cleaning process is dependent upon the chemical activity of the emulsion cleaner as well as the degree of mechanical energy delivered by the spray. Because the mechanical energy facilitates the process of cleaning, the arrangement and design of the spray units are of considerable importance.

Reservoir tanks should have a volume at least three times the capacity of the pump to be used, thus allowing ample time for settling of soils.

Emulsion cleaners, because of a possible fire hazard, should be kept away from heat treating, welding, or similar operations in the plant. Appropriate ventilation must also be provided to comply with health and safety regulations.

14.4 SOLVENT CLEANING

Solvent cleaning is a process that uses common organic solvents, or combinations, as soak cleaners. Cleaning is performed at or slightly above room temperature, with or without agitation.

Solvent cleaning is to be distinguished from vapor degreasing in that, although similar solvents may be used, in vapor degreasing the process is carried out through exposure of the part to solvent vapor, and sometimes by immersion in the hot solvent. Drying is accomplished by evaporation of the solvent from the parts while they are suspended in the hot vapor. In solvent cleaning, parts are dried either at room temperature or by the use of an external heat source.

14.4.1 Composition

Aliphatic petroleum solvents, chlorinated hydrocarbons, Stoddard solvents, and mineral spirits are the most widely used solvents. Commonly, a mixture of a chlorinated solvent with a mineral spirit may be used to take advantage of the excellent cleaning characteristics of chlorinated hydrocarbons and the low cost of mineral spirits.

14.4.2 Equipment and Methods

As is the case with most cleaning operations, agitation increases the effectiveness of the cleaner. With continued use, the solvent becomes contaminated with soil and at some point must be replaced. However, the contaminated solvent can be reclaimed through distillation. The distillation equipment can be operated within the plant if usage justifies the capital expenditure, or contaminated solvents may be sent to companies that specialize in reclamation.

The cleaning equipment line consists of tanks with appropriate mixing devices and with ventilation systems installed to carry away the fumes, which may be toxic and/or flammable. Solvent cleaning is particularly useful as a technique for inexpensive precleaning of parts, thus reducing the cost of subsequent cleaning operations and/or for cleaning of parts with size or configuration that makes them difficult to clean by other methods.

Generally, the process is applied to short-run rather than high-production applications, and because of a lack of uniformity of the cleaning process and the subsequent redeposition of soil on the workpiece, cleaning is relatively less effective. The amount of soil that remains on the part depends upon the initial concentration, how long the solvent has been used, and the degree to which it is contaminated. Frequently, two separate baths are used to provide more uniform cleaning. Because of the low temperatures that are most frequently used, a much wider range of solvent types can be used, thus increasing the effectiveness of the method in cleaning a wide range of soils.

Fire and toxicity are the greatest hazards associated with the use of solvents. Flammable solvents must be stored in appropriate containers, and adequate ventilation should be provided. Operators should not expose their skin to the solvents, and appropriate work habits should be enforced.

14.5 VAPOR DEGREASING

The use of hot vapors of chlorinated solvents to remove oils, greases, and waxes is broadly applied in manufacturing operations when the surfaces of parts are contaminated with mill oils, lubricants, rust preventatives, and cleaning residues.

In general, a vapor degreasing unit consists of a steel tank with a heated solvent reservoir or sump at the bottom and a cooling zone near the top. Between these is a solvent layer, which consists of the hot solvent vapor and does the cleaning. The hot vapor, which displaces the air, condenses when it reaches the cooling zone, thus maintaining a fixed vapor level. The temperature differential between the hot vapor and the cool workpiece entering the cleaning zone causes the vapor to condense on the workpiece, dissolving the surface soil.

14.5.1 Composition

Four principal solvents are used for vapor degreasing: trichloroethylene, perchloroethylene, 111 trichloroethane, and methylene chloride. Their selection depends upon a value analysis of direct costs, recoverability, stability, efficiency of soil removal, water spotting, sensitivity to acid, part shape and size, nonflammability, temperature, and toxicity.

Trichloroethylene has, for many years, been the major vapor degreasing solvent; however, because of air pollution regulations, it has been replaced by other solvents, primarily 111 trichloroethane. Trichloroethylene does not adversely affect common metals (iron, steel, aluminum, copper, and brass) and has excellent solvency. However, it has a tendency to leave water spots, is unstable with respect to water contamination, and is a nonexempt solvent as defined in applicable air-purity standards.

Perchloroethylene is also nonexempt; nevertheless, with good practice, it continues to be used because of its excellent stability, resistance to chemical decomposition, and the ease with which it removes high-melting-point waxes. Because of the high boiling point of perchloroethylene, parts taken directly from the cleaner are too hot to handle, and thus the work cycle must be adjusted accordingly.

111 Trichloroethane is an exempt solvent and has properties similar to trichloroethylene. It has poor water-spotting characteristics, is not stable with respect to caustics, and, in general, cannot be used with water-soluble metalworking fluids because of water contamination resulting in the production of hydrochloric acid.

Methylene chloride, which is also exempt, is effective in removing many oils, fat, greases, and waxes. Its lower boiling point makes it suitable for cleaning parts that may be sensitive to temperature. The product has good general stability, and parts are cool enough to be handled immediately. However, it is not effective in cleaning water-soluble lubricants. This solvent has not been widely used.

Chlorinated solvents may be adversely affected because of the free fatty acid that may be present in certain cutting oils, or because of the presence of sulfur or chlorine also present in many metalworking lubricants. Furthermore, the buildup of oils and greases in the boiling or vapor chamber can cause foaming and a reduction in evaporation. Because cleaning takes place in the vapor phase, contamination of the part surface does not occur as it does in solvent cleaning, where soil builds up in the solvent with use. Nevertheless, when the solvent used in vapor degreasing has a concentration of oil on the order of 25%, it, too, should be replaced, with the oily solvent reclaimed.

Although chlorinated degreasing solvents are inhibited against the effects of hydrochloric acid formation in the presence of water, water should be removed from the degreaser. Most degreasers are equipped with one or more water separators that remove pure water from the circulating recondensed solvent.

14.5.2 Equipment and Methods

In general, a vapor degreaser provides for liquid solvent holding, a heating system to boil the solvent, and a condenser system to recondense or control the solvent vapor level within the equipment.

Vapor phase cleaning is generally the simplest procedure and relies upon the vapor condensate on the workpiece to dissolve the soil. The soil is removed in the liquid phase as it condenses and drips off the part. Workpieces are dry when they leave the equipment.

In more difficult cleaning operations, that is, those in which the workpiece has a complex geometry, a cycle may be used consisting of vapor cleaning. The workpiece first passes through a vapor zone, where preliminary cleaning takes place; then warm solvent is sprayed to reach the areas of the part most difficult to reach during the vapor cleaning step. Subsequently, a second vapor phase cleaning cycle takes place for final cleaning.

In the case of very thin sections, in which the part may be heated by the vapor prior to effective condensation and cleaning, a cycle consisting of vapor cleaning followed by condensation is then followed by a warm liquid dip in which agitation of the workpiece completes the cleaning cycle.

When soils are difficult to clean, a cycle consisting of boiling solvent, followed by a warm solvent rinse, followed by vapor cleaning is used. The violent boiling action in the first stage acts as a scrub, removing much of the heavy lubricant deposits.

14.6 ACID CLEANING

Acid cleaning should be distinguished from pickling because of its relative mildness. Various mineral acids, salts, organic acids, and other organic compound mixtures can be effectively used for removing grease, oil, and various metalworking lubricants. There are limitations to acid cleaning, depending on soil chemistry and the possibility of etching the cleaned metal surface. If oil or chemical deposits are very heavy, acid cleaning is not effective. Acid cleaning should be followed by a rinse and then alkaline cleaning to neutralize any acid residue left on the surface of the part.

14.6.1 Composition

The common acids (H_2SO_4, HNO_3, and HCl) are not effective cleaners for metalworking lubricants or other mill oils and greases, although they are often the acids preferred for pickling. In combination with solvents or salts, surfactants, inhibitors, and organic acids, however, acid cleaning may be used, particularly when oxide is to be removed along with lubricant films.

Phosphoric acid and ethylene glycol monobutyl ether mixtures are used to remove oils and light rust from steel. At the same time, a phosphate coating may be deposited, providing rust protection as well as a base for subsequent painting.

Organic acids, such as citric acid, tartaric acid, or glucosic acid, in combination with acid salts, such as sodium acid sulfate or ammonium persulfate, can be effective cleaners for many metal surfaces.

Solvents, such as glycol ethers, with detergents, such as polyether alcohols, are often used with organic or inorganic acids and may include defoamers and inhibitors, depending upon the characteristics of the cleaner as well as soil to be removed.

14.6.2 Equipment and Methods

Acid cleaning is preferred when parts are covered with heavy paste or oily deposits and are rusted. Organic acids are used when corrosivity of the metal surface must be minimized and the higher costs of the organic acids can be justified. In the case of aluminum cleaning, chromic acid and sulfuric acid mixtures may be used to remove surface oxide and oxidized oil films. The acid cleaners remove surface oxides and light chemical coatings on aluminum resulting from lubricant deposits.

Spray cleaning for high-production rates, immersion cleaning (the most versatile process if racking of the part is economical and practical), barrel cleaning for large quantities of small formed parts, or hand wiping for low-production complexly shaped or large parts, are the processes used for the majority of acid cleaning. Electrolytically assisted cleaning may be used if lubricant surface coatings are very light, parts are regular and simple in shape, and time for the cleaning cycle is to be reduced.

14.7 TEST PROCEDURES

The selection and control of an appropriate cleaning method can be carried out through the use of the following relatively simple test procedures.

1. Waterbreak test. The metal with a typical soil residue is dipped into a mild acid bath, followed by a dip in clean water. If the surface is not continuously coated by the water, the surface is not clean.
2. Neilson method. Ten panels are cleaned, measuring the time required to obtain visually similar results. Panels are checked by the waterbreak test and then by an acid copper test, which consists of immersion in a copper sulfate solution (typically 5 oz copper sulfate and 1 fluid oz sulfuric acid per gallon of water). On a clean surface, copper is deposited as a semi-bright coating that is free of spots.
3. Wiping method. This is a qualitative test that consists of taking a cleaned panel, wiping it with a white cloth or paper, and examining the paper or cloth for any residue that reflects poor cleaning.

Various other methods have been developed for carrying out the evaluation of cleaning effectiveness. Simple test methods are often sufficient.

BIBLIOGRAPHY

Metals Handbook, vol. 5, 9th ed., 1982. American Society for Metals, Metals Park, Ohio, p. 5.

Rosen, M. C., 1978. *Surfactants and Interfacial Phenomena,* Wiley, New York, Chapters 8, 9, and 10, pp. 224–291.

15
Federal, State, and Local Laws and Regulations

The interface between private enterprise and public welfare is one that is characterized by considerable friction in constantly changing positions. The manufacturer, the user, and the public have ill-defined, but nevertheless high stakes in the adoption and enforcement of laws and regulations governing the operations of these several interests. Some of the laws and regulations that have particular impact on metalworking lubricant development, manufacture, use, and disposal are discussed in this chapter.

In considering the operation of the government, it is useful to appreciate the incomplete nature of the information upon which present regulatory activities are based. Nevertheless, since societal concerns about jobs often clash with society's demand for a risk-free environment in which to work and play, laws and regulations will be enacted and compliance will be required with widely varying interpretations of what is mandatory.

There are several hundred federal, state, and local laws and regulations that prescribe the manufacture, use, and disposal of metalworking lubricants. Many of these deal with pricing, antitrust, employment, and other areas. Only some of the applicable laws and regulations dealing with the technical concerns relevant to metalworking lubrication are discussed here.

15.1 THE ENVIRONMENTAL PROTECTION AGENCY AND THE TOXIC SUBSTANCES CONTROL ACT

The Toxic Substances Control Act (TSCA) of 1976 delegates to the Environmental Protection Agency (EPA) the authority to regulate the manufacture, processing, distribution, use, and disposal of chemicals to avoid unreasonable risk to health or to the environment. The TSCA program consists of essentially two different operations. One involves the cataloging of all commercially distributed substances or mixtures and their use, evaluation of the potential risk associated with their use, and regulation of those substances and uses that pose a risk. The second operation involves the regulation of new chemical substances or mixtures.

A company interested in commercializing a new chemical, that is, a chemical that has not been cleared through the first list, must submit a premanufacture notice form (PMN) to the EPA on any new composition if it is subject to TSCA regulation. The decision on whether the chemical or composition is subject to TSCA regulation depends upon the nature of the mixture and the presence or absence of chemical reactions during manufacture or use of the material in question. Mixtures and certain polymers are not subject to registration. A bonafide search and telephone inquiry normally precedes a PMN.

Obviously, this screening of potentially hazardous materials involves questions of economic value that must be answered prior to a decision about use. If a lubricant consists of components that do not undergo chemical reaction and that have been cleared for use previously by the EPA, there are no additional costs in complying with TSCA regulations. On the other hand, if either of these conditions has not been met, then additional costs of testing and filing a PMN prior to manufacture and use must be taken into account in making any business judgment about the commercialization of a new lubricant.

15.2 THE ENVIRONMENTAL PROTECTION AGENCY AND THE CLEAN AIR ACT

The Clean Air Act of 1970, and particularly the 1977 amendments thereto, involve questions of atmospheric contamination that directly relate to the admissibility of use of a given lubricant. The problem is the emission of volatile components or decomposition during a metalworking operation with a particular lubricant. As is the case with much environmental regulation enforcement, there remain many questions about the regulations that have been adopted or are to be adopted.

Nevertheless, the use of lubricants in a plant, which volatilize during the metalworking operation, or at any time for that matter, involves consideration of the possible hazardous nature of these volatile components. Most solvents that may be used as components of metalworking lubricants, for example, have permissible concentrations in the air that must not be exceeded. Nevertheless, the producer as well as the user of metalworking lubricants should be forewarned of possible Clean Air Act violations that may be attributable to a particular lubricant, Similarly, if the lubricant in question decomposes during metalworking and the decomposition products volatilize, the same set of questions must be satisfactorily answered to proceed with commercialization and use.

15.3 THE ENVIRONMENTAL PROTECTION AGENCY AND THE WATER POLLUTION CONTROL ACT

The Water Pollution Control Act of 1972, as amended, requires that every source of waste disposal that may enter a waterway must be identified with a governmental permit that sets out the conditions and types of wastes allowed to pass into the water.

The planning program requires that each state develop a comprehensive plan to upgrade the quality of all waterways in the state to meet federal standards. As these developments occur, manufacturers of lubricants and their customers can expect that there will be increasing restrictions on the conditions and types of pollutants that can be disposed of into waterways under the permit system.

In time, disposal of industrial waste into waterways will probably be completely prohibited. Furthermore, under the clean water act, any industry that discharges its waste into a municipal treatment system must pretreat its effluent so that it does not interfere with the municipal water treatment operation. The EPA publishes a list of toxic pollutants each year, as authorized by law. Each of these pollutants is subject to effluent limitations using the best available technology for cleanup.

15.4 THE ENVIRONMENTAL PROTECTION AGENCY AND THE SAFE DRINKING WATER ACT

The Safe Drinking Water Act of 1974 provides for the protection of drinking supplies from contamination by toxic wastes through the establishment of national drinking water quality standards. The Act calls for studies of contamination by cancer-causing chemicals, a task in which the EPA laboratories play a major role.

15.5 THE ENVIRONMENTAL PROTECTION AGENCY AND THE RESOURCE CONSERVATION AND RECOVERY ACT

The Resource Conservation and Recovery Act (RCRA) requires the EPA to establish standards and regulations for hazardous waste management and land disposal. Criteria for the definition of a hazardous waste are to be established, and generators and transporters must report on the character of the wastes generated and the manner in which they are transported. Furthermore, the agency is to develop and promulgate regulations for the storage, treatment, and disposal of hazardous wastes. The EPA Must issue criteria for acceptable land disposal for all those wastes that do not fall within any of these classifications. Therefore, disposal of waste oils is governed by appropriate regulations.

15.6 THE OCCUPATIONAL SAFETY AND HEALTH ADMINISTRATION

Along with the EPA and its responsibilities for the protection of clean air and water, the establishement of the Occupational Safety and Health Act (OSHA), which concerns itself with unsafe conditions in the workplace, regulates all aspects of the manufacture, transportation, use, and disposal of metalworking lubricants.

The basic purpose of the law establishing OSHA is to encourage employers and employees to reduce hazards in the workplace and to implement new or improved existing safety and health programs. Additional aspects of the law are to promulgate regulations that mandate reporting and record-keeping procedures, monitor job-related injuries and illnesses, and develop job safety and health standards and enforce them effectively. The states are encouraged to assume the fullest responsibility for establishing and administering their own occupational safety and health programs, which must be at least as effective as the federal program.

To carry out its responsibilities under the Act, OSHA is responsible for promulgating legally enforceable standards. The National Institute for Occupational Safety and Health (NIOSH) was established by the Act as an agency of the Department of Health and Human Services. NIOSH conducts research on various safety and health problems and recommends standards for adoption by OSHA. Especially important is their investigation of toxic substances and development of criteria for the use of such substances in the workplace.

The establishment of standards has, for example, led to an investigation of the formation of nitrosoamines in cutting fluids. After several years of investigation, standards have yet to be established because of the complexity of the problem involved in relating the used cutting fluid to the specific nitrosoamine and its supposed hazardous nature. Not only is the relationship between particular lubricants and the work environment extremely difficult to relate to specific hazards, but the analytical methods that are available leave much to be desired. However, recently a PMN was rejected on the basis of possible additions of nitrites resulting in the formation of nitrosoamines.

The Occupational Safety and Health Administration, operating under the Department of Labor, can impose punishments upon violators of regulations of up to $20,000 and/or imprisonment for 1 year. Obviously, no employer is likely to allow the use of a lubricant that may result in conflict with the OSHA standards. The lubricant formulator should consider the chemistry of the formulated lubricant and its possible toxicological effects. Furthermore, combustibility or explosiveness of the lubricant must also be considered, and any undesirable reaction products must be recognized and provided for.

Specific and detailed OSHA standards, where they exist, must be taken into account in the formulation of metalworking lubricants. The establishment of standards can and should be influenced not only by the worker and employer, but also by the developer of the metalworking fluid. All proposed standards are announced in an Advanced Notice of Proposed Rulemaking as set out in the Federal Register. The Advanced Notice sets out the working conditions to be considered, the reasons standards are necessary, and suggested standards. Interested parties should respond to this notice. After a specific time period, during which the public and interested parties have an opportunity to present information pertinent to the proposed standards, permanent standards may be established.

In formulating a lubricant, an awareness of and sensitivity to OSHA regulations will not only minimize the cost of compliance but may also lead to improved formulation. For example, will there be skin sensitivity or eye irritation potentially resulting from repeated contact with the lubricant? Are any volatile or combustion products given off by the lubricant? Will undesirable reactions with the metal surface take place? What effects may tramp oil contaminants have on the reactivity of the lubricant? Will the allowable concentration and time exposures for air contaminants be exceeded?

15.7 THE NATIONAL INSTITUTE FOR OCCUPATIONAL SAFETY AND HEALTH

The same act that established OSHA also established NIOSH. The Institute was administratively placed in the Department of Health and Human Services. It is the responsibility of NIOSH to carry out research, experiments, and demonstrations that identify toxic materials, determine safe exposure levels, and develop analytical methodology and screening methods for establishing health hazards. Furthermore, NIOSH develops and recommends occupational and health standards, and if the standards concern the general public, EPA is responsible for implementing them. If the standards relate to the workplace, OSHA has this responsibility.

Of particular impact on the formulator of lubricants is a list of approximately 1500 chemicals that have been suspected as carcenogenic and are on a list for investigation by NIOSH. Because of the complexity of the issues involved, only about 19 potential carcenogens have been regulated to date by appropriate OSHA standard formulation.

In summary, it is the responsibility of OSHA to prevent occupational injury, which includes both acute and chronic illness caused by inhalation, ingestation, or contact with toxic substances. Companies are required to keep records of injury or illness. These logs must be posted in the plant each year for the whole month of February. OSHA has the right to inspect any workplace. NIOSH is the investigative arm with the charge to present facts and recommendations that may lead to the establishment of OSHA standards.

15.8 PRODUCT LIABILITY

The implications of the law regarding product liability have been so broadened with respect to their impact on the manufacturre of industrial lubricants that no one engaged in either the use or manufacture of lubricants can afford to ignore the financial as well as liability implications.

To illustrate the breadth of the financial and legal implications of product liability, let us use the example of a manufacturer of lubricants that formulates a water-soluble cutting fluid that is an emulsion, which for the purposes of this discussion contains a chlorinated paraffin. Let us further assume that a machine operator develops dermatitis, for reasons not clearly understood but related to that individual being found, in subsequent patch tests, to be sensitive to chlorinated paraffin. Let us further assume that because of the dermatitis, acute in this case, the worker scrapes a pustule (caused by dermatitis) while adjusting a tool. It is painful, and involuntary motion has brought the hand in contact with the tool and, as a consequence, a finger is severed. The law that covers the responsibility of a manufacturer to produce a product that is defect free also includes the responsibility to design a product that is appropriate for the use intended. Thus, the lubricant manufacturer could be held responsible for including a chemical in the formulation that resulted in the dermatitis, which in turn resulted in unsafe practices by the worker. Furthermore, the plant management could be held responsible for this same injury because of its negligence in protecting the worker by seeing to it that the individual no longer worked in an environment injurious to health. Thus, the liability would extend from the user to the supervisors and to the suppliers. Furthermore, not only would the supplier of the lubricant be potentially liable, the supplier of the chlorinated paraffin to the lubricant manufacturer could also be held liable.

Although all this requires that the victim prove that the party being sued should have been more careful, and that this lack of care caused the victim's injuries, recent developments have led to the adoption of a concept called "strict liability." This makes the manufacturer of lubricants liable for injury to others even if it cannot be proven that the firm was negligent. Apparently, the court has decided that those who supply the lubricants have a responsibility to carry sufficient insurance to compensate anyone injured by the product no matter how far removed the injury from the original lubricant use decision.

It now becomes imperative, because of the high cost of product liability, more than ever to be aware of the possible hazards involved in the use of any given lubricant formulation. Furthermore, the lubricant supplier has the responsibility for identifying possible hazardous operating conditions under which the lubricant may be used. If there is consideration only of the functional operation of the lubricant, and there is not sufficient understanding of the user's operating parameters, the lubricant supplier, equipment producer, and the user may be liable for costly damage claims as well as harmful publicity.

There are, of course, other regulations that have an impact on the research and development activities associated with the technological development of improved metalworking lubricants. Patent and trademark regulations may directly influence the protection granted the inventor of a new lubricant composition or chemistry, as well as the commercial identification of the lubricant.

Investigations of the effects of nitrosoamines in metal cutting fluids has resulted in proposed definitive government regulations. The case of nitrosoamines is a good example of the complexity associated with ensuring zero risk in an industrialized society, since nitrosoamines apparently are also present as a result of food digestion.

Furthermore, in the past few years, attention has been devoted to a requirement for labeling containers holding lubricants, to reveal their composition to the worker or to anyone else to whom such information might be of interest. The impact on future developments of technologically advanced compositions is difficult to predict, but it certainly could be far reaching. Who would wish to invest in the development of a new formulation only to see its composition become common knowledge with shipment of the first container purchased?

In this brief review, we have barely touched the full impact of national and local laws and regulations as they influence the development, manufacture, and use of metalworking lubricants. It is probably impossible to realistically evaluate the influence of this legal and regulatory environment on the various interested parties. For the purposes of this chapter, it is important to appreciate that the influence is pervasive and must not be ignored in considering lubrication of metalworking operations.

BIBLIOGRAPHY

Baler, E. J., 1982. *American Industrial Hygiene Association Journal,* 43(9), Do occupational health guidelines and standards protect workers?.

Crotty, J. J., 1982. *American Industrial Hygiene Association Journal,* 43(9), The impact of government regulations on the use of cutting fluids, 7-9.

Ehrlich, W., 1982. Living with toxicity regulations, *NLGI Spokesman,* July, 120-126.

Federal Register, vol 49, no. 15, Monday, January 23, 1984, Rules and Regulations.

Opalchik, M., ed., 1982. *Industrial Fluids, Controls, Concerns, and Costs; RCRA's Impact on the Metal Working Industry,* Society of Manufacturing Engineers, Dearborn, Michigan, pp. 3-6.

Smith, M. D., and C. H. Wist, 1969. How plant practices affect employees health in the presence of metal working fluids, *Journal of the American Society of Lubricating Engineers,* August, 321-325.

Sontag, J. M., ed., 1981. *Carcinogens in Industry and the Environment,* Marcel Dekker, New York, pp. 168-281.

16

Selection of Lubricants for Metal Forming Processes

Industrial lubrication practices for specific groups of metalworking processes are described in this chapter, together with general recommendations for the types of lubricants suitable for these processes. Because of the wide variety of lubricants available and the wide range of process and material variables in metalworking operations, the selection of appropriate lubricants and lubrication practices are difficult to categorize. Thus, only general guidelines can be given within the scope of this volume.

In the sections that follow, individual groups of metal forming processes are outlined in terms of the necessity of lubrication, methods of application, and critical aspects of lubrication unique to that process, together with generally recommended lubricants. Reference to the materials covered in Chaps. 2, 4, and 6 should also be made to gain a broader perspective of the subject matter of this chapter and to appreciate the reasons behind the selection of a particular lubricant for a specific process.

16.1 FORGING

Forging may be carried out at room temperature, or at medium or high temperatures. The operation can be performed in an open die, impression die, or closed die. By its nature, forging is basically a batch process and lubrication is generally necessary for one or more of the following purposes:

1. Reduce forging load
2. Improve material flow in the dies

3. Prevent die seizure
4. Reduce die wear
5. Act as a thermal barrier between the workpiece and die surfaces
6. Help release the forged part from the dies

In addition to the foregoing characteristics, forging lubricants, as in other lubricants, should also have the following properties:

1. Wet the surfaces uniformly to avoid local breakdown of lubrication
2. Be free of residues that might accumulate in deep recesses of forging dies to affect tolerances and/or part definition; neither should it deposit on equipment and surroundings and be difficult to remove
3. Not corrode the dies, but should provide a protective coating
4. Provide the proper amount of cooling
5. Be applicable automatically, preferably by spraying
6. Should not smoke, burn, pollute the air, or become a health hazard
7. Should not separate while standing
8. Concentrate should be dilutable to permit a range of lubricating characteristics
9. Be cost effective

16.1.1 Lubrication Regimes in Forging

With liquid lubricants, lubrication in forging is generally of the mixed-film type; that is, the contact forces are supported partly by a fluid film and partly by metal-to-metal contact. Solid lubricants are also commonly used in forging, in which case lubrication is provided by a solid film with its inherently low friction, although the oil or water used as a carrier may also influence the film strength of the lubricant.

The geometry of the die-workpiece interface in forging is such that thick films are generally not formed, except in simple upsetting operations at high speeds or in isolated cases in which the contact geometry, viscosity of the lubricant, and the relative sliding speed are favorable for thick films to form.

As in all metalworking processes, the choice of an appropriate forging lubricant depends on many factors. Among the more important considerations are the forging temperature, speed, and die surface finish.

16.1.2 Cold Forging

For forging at room temperature (which may also be listed under cold extrusion because of the similarity of metal flow) lubricants are generally compounded, using higher viscosity fluids if lower deformation rates are involved, such as is the case when upsetters or presses are used. If, on the other hand, high impact velocities are involved, such as in high-energy-rate forging, then hydrodynamic lubrication can take place. A low-viscosity lubricant may be appropriate in this case.

In cold-extrusion operations, which are among the most severe forging operations and involve very high contact pressures, a two-phase lubricating system is generally used, consisting first of a solid coating of a phosphate or oxalate, followed by a lubricant film of soap. Typically, lime or phosphate coatings are applied to carbon or alloy steel, and oxalate on stainless steels.

16.1.3 Warm Forging

Warm forging, carried out at temperatures ranging from 400 to 600°F (200 to 300°C) for aluminum alloys and 900 to 1300°F (500 to 700°C) for nickel-based alloys, is attractive because satisfactory part finish and closer tolerances can be obtained at greatly reduced die pressures. Because of the temperatures involved, most organic lubricants are ineffective and must be replaced by solid lubricants, such as graphite or inorganic salts.

16.1.4 Hot Forging

Temperatures in hot forging may range from 750°F (400°C) for aluminum alloys to about 3000°F (1650°C) for refractory metal alloys. The lubricants used are commonly oil or water based, which serve as carriers for solid lubricants, such as graphite and molybdenum disulfide and are generally applied on the dies. A reduction in die temperatures enhances die life. However, care in maintaining a constant temperature is equally important. Wide fluctuations in temperature can lead to die failure.

Because of their detrimental environmental effects, oil-based lubricants are now being replaced by water-based fluids. Glass coatings are also used for hot forging of nickel-based alloys, refractory alloys, and titanium, both as a lubricant and also to prevent oxidation of the workpiece. In the case of copper and steel, the oxide that forms on the surface of the workpiece at high temperatures may also aid in lubrication because of die insulation, as well as possibly lower friction due to the presence of an oxide layer. Jacketing (canning) of the preform may also be practiced in the case of reactive metals or in forging powder-metal preforms.

16.1.5 Rotary Swaging

The use of lubricants in rotary swaging can have a significant influence on surface finish and die life. This process is essentially a form of cold forging and utilizes a lubricant film to prolong the life of swaging dies and improve surface finish. However, the lubricant can also have an adverse effect on the feeding of the workpiece through the swaging dies, particularly in the case of dies with large angles, since in swaging the part is pushed through the dies. This problem is less severe with powerful automated swagers than in manual feeding.

If the workpiece is tubular in shape, a lubricant at the mandrel-tube interface is essential to prevent seizure and galling. A wide variety of liquid and solid lubricants, and their combinations, can be used successfully for mandrel lubrication in rotary swaging.

16.1.6 Application Methods

The application methods for lubricants in forging consist primarily of spraying or swabbing. Solid layer-lattice lubricants (graphite or molybdenum disulfide) may also be applied to the die surfaces by bonding with a polymer. It has been observed that, in many cases, the method of lubrication and its uniformity at the die-workpiece interfaces can be more important than the type of lubricant chosen.

Automatic spraying equipment has been developed that applies the lubricant uniformly and efficiently. Proper treatment and coating of die surfaces with lubricants are important aspects of forging lubrication.

16.1.7 Recommended Forging Lubricants

Lubricants generally recommended for cold and hot forging of various groups of materials and their alloys are given in Table 16.1 It should be noted that many lubricants are often modified by various additives for special applications.

A number of lubricants for metalworking processes are being investigated and used either in industrial plants or in research conducted at various institutions. Coverage of these experimental developments are beyond the scope of this volume.

16.2 ROLLING

Rolling is an important metalworking process, which, unlike many deformation processes, cannot be carried out without friction. Except for initial breakdown passes, rolling is a continuous operation, and the need for good surface finish, especially in cold rolling, is an important consideration.

Lubrication in rolling primarily serves the function of reducing friction and roll wear, although minimum friction is necessary for the operation to

Table 16.1 Lubricants for Forging[a]

Material and its alloys	Cold	Hot
Aluminum	MO + FA, S, CC + S	G + MO, G + W, G + binder, S, MoS_2
Beryllium	MO + FEs	MoS_2, J
Copper	S, E, T	G + W, S
Magnesium	MO + FC, CC + S	G + W, MoS_2 + G
Nickel	–	MoS_2 + W, G + W, Gl
Refractory metals	–	Gl, G, J
Steels		
Carbon; low alloy	MO + EP, CC + S	G + W, G + MO, Sa, Gl
Stainless	CP + MO, CC + S	G + W, G + MO, Gl
Titanium	CC + S, MO, various coatings (polymers, oxides, metals) + lubricant	G + W, MoS_2, Gl

[a]Codes: D – dry; E – emulsions; F – fat; G – graphite; Gl – glass; S – soap; T – tallow; W – water; J – jacketing (canning); MO – mineral oil; FA – fatty acids; FC – fatty compounds; P – polymer; Es – esters; CC – conversion coating; Wa – wax; Cl – chlorine additives; EP – extreme-pressure additives; MoS_2 – molybdenum disulfide; Pa – paraffin; CO – compounded oil; Sa – salts; FO – fatty oil; L – lanolin; CP – chlorinated paraffin.

take place. Depending on material and process parameters, temperature control is another important function of lubrication in rolling. Excessive heat generation (due to plastic work of deformation and friction) can cause variations in the diameter of the rolls, hence in the dimensions of the rolled strip. Roll wear is influenced not only by contact and frictional stresses but also by temperature. As in forging dies, roll wear may take place by various mechanisms, such as adhesive, abrasive, and thermal fatigue wear.

16.2.1 Cold Rolling

Cold rolling, which is typically carried on to produce strip, is performed at high speeds, with reductions of as much as 50% per pass. The type of lubrication generally obtained in cold rolling is of the mixed-film regime. Depending on process parameters, such as rolling speed, roll diameter, and reduction per pass, hydrodynamic films may also be developed at the roll-workpiece interfaces. These films are, however, rather unstable, and full fluid-film lubrication in cold rolling does not generally represent common rolling practice.

A number of theoretical studies are available analyzing the complex interrelationship among effects of temperature, roll flattening, lubricant characteristics, and various process parameters. As the roll pressure increases, the influence of a boundary film predominates. Under sufficiently severe conditions, the lubricant film may break down and severe roll pickup may occur (particularly with aluminum, steels, and stainless steels), resulting in poor surface finish of the rolled product.

The choice of rolling lubricants depends not only on the workpiece material but also on whether lubrication or cooling is the desired controlling mechanism. Generally, the most commonly used lubricants are oil or water based, with various additives to improve film strength. As in forging lubrication, the trend is to use water based fluids to reduce environmental contamination (smoke and particulates) and provide the desired cooling. In cold rolling of nonferrous alloys, staining of the strip by the lubricant must be avoided by appropriate process control and choice of lubricant.

16.2.2 Hot Rolling

In hot rolling, in which the temperature range is similar to that in forging, mixed-film lubrication typically occurs. Rolling can also be carried out dry, in which case the oxides on strip surfaces are effective as solid lubricants, particularly in rolling steels and titanium. For refractory alloys special techniques, such as jacketing, are used.

Effective lubrication in hot rolling reduces roll wear significantly. In hot rolling of steel the effect of lubrication is not as important as that in rolling aluminum. In the latter case, if rolled dry, the rolls may be rapidly coated with aluminum and its oxides and rolling cannot be carried out properly.

16.2.3 Shape Rolling

Lubrication is an important aspect of shape rolling because of the high relative sliding at the roll-workpiece interfaces and the resulting nonuniform wear of the rolls. Although lubrication practices are similar to those in flat rolling, in shape rolling it is necessary that lubricant supply be adequate in critical regions, and that film strength is sufficient to resist the wiping action of the rolls. Tube rolling involves similar considerations.

Lubricants for hot shape rolling and tube rolling are suspensions of graphite, salt solutions, or other lubricants similar to those developed for hot

Table 16.2 Lubricants for Rolling[a]

Material and its alloys	Cold	Hot
Aluminum	FA + MO, MO, polymer emulsion + synthetic solutions, MO + FC	FA + CO, FA + E
Beryllium	–	G suspension
Copper	E, MO + FA, S	D, E
Magnesium	D, FA + MO, polymer suspension, synthctic solutions	MO, D, G suspension
Nickel	MO + Cl, P + F, E	G, Sa, or G suspension in W
Refractory metals	MO, P + Es	Gl + MoS_2, MO + EP
Steels		
Carbon; low alloy	E, CO, synthetic solutions	W, MO, CO, G + E
Stainless	CP + E, Cl Pa + MO	D, MO + Cl additives
Titanium	MO, CC + FO, G, Sa solutions	P, CC, G, MoS_2

[a]See Table 16.1 for explanation of abbreviations.

rolling. Cold rolling of shapes and tubes generally utilize cold rolling lubricants. The control of lubricant flow and coverage is very important in these processes.

16.2.4 Application Techniques

Application techniques typically involve spraying the rolls uniformly with the lubricant or coolant (such as water) using nozzles placed along the length of the rolls. Nozzle location is often determined by the need for camber control of the roll. Solid or semisolid lubricants, such as graphite or tallow formed into solid blocks, may be applied to the surface of the roll under light pressure. As they coat the roll surface these blocks wear away until they are replaced.

16.2.5 Recommended Rolling Lubricants

The recommended lubricants for cold and hot rolling of various metals and their alloys are listed in Table 16.2. The general comments made in Sec. 16.1.1 also apply to this and subsequent sections in this chapter.

16.3 EXTRUSION

Extrusion is a batch process and may be performed in a variety of ways, such as by direct, indirect, or hydrostatic extrusion techniques. For hollow extrusions, special dies (bridge, porthole, or spider dies) are used. The role of lubrication in extrusion depends on the particular process.

Generally, lubrication in extrusion reduces frictional forces and permits the metal to flow more homogeneously. Furthermore, with effective lubrication, die life is improved and temperature rise is lower. Excessive temperature rise can cause cracking (hot shortness) or incipient melting of the extruded product. In direct extrusion, where the billet moves toward the die and thus involves friction along the chamber wall as well as in the die, lubrication can have a significant influence on material flow and forces. In indirect extrusion, on the other hand, there is no relative motion at the billet-container interface. In hydrostatic extrusion, the fluid used in the chamber also serves as a supply of lubricant at the die.

Although lubrication is essential in most extrusion processes, excessive lubrication can be detrimental to the surface finish of the product since the surfaces produced do not undergo any burnishing at the die-workpiece interface. In certain cases lubrication may be eliminated, or it may be unnecessary. For instance, with square or flat dies (90° die angle, used with softer nonferrous metals), a dead metal zone develops and the material flows along (shears) at a 45° angle and hence the billet does not contact the square die. Instead, the material shears within itself. Also, for billets with oxidized surfaces, the ram may be made a little smaller than the chamber diameter and, during extrusion, a thin "skull" is left in the chamber consisting of the oxidized surface, which is then discarded.

In extruding hollow shapes with porthole or bridge dies (used primarily with aluminum), no lubricant should be present in the extrusion zone; otherwise, rewelding of the material at the die exit is not possible. Furthermore, in situations in which chevron cracking (also known as centerburst or arrowhead cracking) occurs, for example, due to small extrusion ratios and/or high die angles, higher friction is generally desirable to minimize the tendency for cracking.

16.3.1 Cold Extrusion

Much of what has been discussed in regard to forging lubrication is also applicable to extrusion. Because of the high room-temperature strength of the workpiece materials, stresses in the tooling are high and thus effective lubrication in cold extrusion is critical. The most widely used lubrication technique,

Table 16.3 Lubricants for Extrusion[a]

Material and its alloys	Cold	Hot
Aluminum	D, G, MoS_2, L, S	D, G, P
Beryllium	–	MoS_2, G, J
Copper	S, T, L, Wa, G, MoS_2	D, MO + G
Magnesium	D, T	D, P
Nickel	CC + S	Gl, J
Refractory metals	CC + S, CC + F	J + Gl, Gl
Steels		
Carbon; low alloy	CC + S, T, MoS_2 + W	Gl, G
Stainless	CC + S, MO, copper plating	Gl
Titanium	CC + G, CC + S	J + Gl, Gl
Zinc	D, T, fatty emulsion	D, T
Zirconium	CC + G, CC + MoS_2	J + Gl + G

[a]See Table 16.1 for explanation of abbreviations.

especially with steels, is phosphate or oxalate conversion coatings, followed by application of soap or wax as a lubricant.

16.3.2 Hot Extrusion

Temperature in hot extrusion range up to about 4000°F (2200°C) for refractory metal alloys. Especially with high-strength materials, hot extrusion can be a difficult operation because of the tendency for adhesion at the interfaces, requiring special die materials and/or coatings and special techniques. Although graphite can be used as a lubricant, the most common lubricant in hot extrusion is glass. It is applied to the hot billet surfaces in powder form, or a cake of molded glass is placed at the front of the billet at the die entry. The viscosity index of the glass cake and extrusion speed may be varied and are important parameters in providing an adequate supply of molten glass lubricant throughout the length of the billet during the extrusion cycle. Commonly used lubricants for extruding various metals are given in Table 16.3.

16.4 DRAWING ROD, TUBE, AND WIRE

Drawing through dies is a continuous, steady-state process for wire drawing; it is, however, a batch operation when drawing bars or tubes on a drawbench. As in other metalworking processes, lubricants reduce drawing forces and power consumption, reduce die wear, and keep temperatures low. Because surface finish is an important aspect of drawn products (since they are generally considered to be finished products), the role of lubrication can be complex. On the one hand, good lubricants should reduce friction by providing, ideally, hydrodynamic lubrication at the workpiece-die interface. On the other hand, such a thick-film regime produces a rough surface finish, depending on the grain size of the metal and viscosity of the lubricant, because the surfaces cannot undergo the burnishing action available from mixed-film or boundary lubrication.

Thus, a thin lubricant film is desirable in drawing, which can be obtained by using low-viscosity lubricants and higher die angles. However, contact pressures at the die-workpiece interfaces can be very high. Consequently, lubricant film strength must be sufficient, such as that obtained by using EP and various other additives, if acceptable die life is to be achieved.

In cases in which internal cracking (chevron) occurs, this can be reduced not only by decreasing the die angle and increasing the reduction per pass, but also by reducing friction.

16.4.1 Application Methods

Two basic methods of lubrication in drawing that are commonly used are dry and wet. In the former, solid or semisolid lubricants (typically soaps) are applied over a lime coating or a conversion coating (phosphate), covering the surface of the material entering the die. In wet drawing, the whole system is immersed in the lubricant (typically oils or emulsions), thus providing ample supply. It should be noted that, in this way, some cooling is also accomplished, depending on the lubricant. Emulsions, typically water based, have a higher cooling capacity than oils. Cooling in drawing can be important; otherwise high temperatures lower the strength of the final product, may promote die pickup, and hence result in inferior surface finish.

16.4.2 Drawing Tubes

Lubrication can be a critical factor in tube drawing, depending on whether the mandrel (plug) is moving or stationary. By analyzing the mechanics of the tube drawing process it can be shown that, in drawing with a stationary mandrel, friction is generally not desirable. On the other hand, with a floating plug, some friction is helpful in reducing the internal stresses in the tube, thus reducing the tendency for failure under longitudinal tensile stresses.

16.4.3 Wire Forming

Drawn wire is formed into many shapes, and lubrication is an important aspect because of the relatively severe sliding between the wire and the dies. Lubricants can reduce the tendency for galling, seizure, and fracture of the wire by reducing the forces involved, as well as contribute to improved die life. Depending on the wire material, lubricants may consist of various oils and emulsions, with EP and other boundary additives, including solid lubricants, such as graphite, molybdenum disulfide, and talc.

The residual lubricants from the wire drawing operation may be sufficient in many cases to serve as a lubricant in wire forming, although additional lubrication may be necessary for severe forming operations.

Removal of the lubricant is important if the formed wire product is to be subjected to subsequent processing, such as plating or coating. Ease of cleaning influences the choice of lubricant chemistry.

16.4.4 Tube Bending

In bending tubes with a mandrel, the outer and inner surfaces of the tube may require different lubricants. The interface between the mandrel and the inside of the tube is generally lubricated with viscous mineral oils or pastes. Special mandrels may also be designed to supply additional lubricants from the inside. The outer surfaces, which are in contact with dies, such as wiper dies, usually require lubricants with low viscosity. This is to prevent the build-up of any heavy coating of lubricant that could interfere with the bending operation and possibly cause wrinkling of the tube.

Since tube forming, especially of nonferrous metals, such as aluminum, copper, and brass, is basically a finishing operation, staining by the lubricant is an important consideration in lubricant selection. In tube bending, because of the pressure and temperature generated during bending on a mandrel, the lubricant film must be tenacious and adhere to the contact surfaces. However, cleaning of the tubes can then be a problem. Appropriate lubricant formulations can be used to minimize cleaning problems.

16.4.5 Lubricants for Drawing

Typical lubricants used in drawing various metals and alloys are given in Table 16.4.

Table 16.4 Lubricants for Drawing Rod and Wire[a]

Material and its alloy	Cold	Hot
Aluminum	MO + F, S, Wa, P suspension, E	–
Beryllium	–	G suspension
Copper	MO + FC, S + Wa, synthetic solutions	–
Magnesium	MO + FC	–
Nickel	CC + S, MO + Cl, P suspension	–
Refractory metals	Cl compounds, phosphate coating + lime + S	G + MoS_2, P
Steels		
Carbon; low alloy	CC + S, S, E (music wire)	–
Stainless	Oxalate coating + S, MO + Cl	–
Titanium	CC + P, CC + S	–

[a]See Table 16.1 for explanation of abbreviations.

16.5 SHEET-METAL FORMING

In bulk-deformation processes, such as forging, rolling, extrusion, and drawing, new surfaces are generated, the extent of which depends on the reduction. The contact pressures are generally high; in certain regions these are multiples of the yield strength of the material.

In contrast to this type of deformation, sheet-metal forming generally involves little or no thickness change, except in such cases as ironing deep-drawn cups. The deformation of the sheet metal takes place typically by biaxial tensile stresses in the plane of the sheet. However, there is considerable relative motion, and the contact area at the die-workpiece interfaces is usually large.

Surface finish is an important aspect of sheet forming because these processes are basically finishing operations. Surface blemishes, such as scratches, scoring, die pickup, and galling, are usually not tolerated.

Surface roughness on both the tooling and the sheet metal can influence lubrication significantly. Some surface roughening can lead to trapping of the lubricant, thus promoting a continuous lubricant supply during deformation of the sheet metal. The importance of effective, consistent, and uniform lubrication to obtain good surface finish as well as promote good shape and tolerance control must be emphasized. Nonuniform lubricant thickness may cause defects, such as wrinkling or folding, particularly with thin workpieces, a phenomenon that may also occur in drawing thin-walled tubes.

A further complication is due to the regions at die-workpiece interfaces that often require preferential lubrication to improve formability. In deep drawing, for instance, it is desirable for the punch to travel with the cup being formed (thus high friction is desirable), whereas in the blank holder region there should be a supply of lubricant sufficient to reduce longitudinal tensile stresses and prevent the cup from being punched out.

There is a wide variety of sheet-forming processes (generally called press-working), ranging from simple bending or stretching operations to fine blanking, deep drawing, and spinning complex shapes. Unlike bulk-deformation processes, temperature rise in sheet-metal forming is not a particularly significant factor, unless the material is of high strength, the deformation is complex, or the workpiece is heated to improve its formability or reduce forces.

Lubrication is more important to fine blanking (fine-edge blanking), in which clearances are very tight. Typical lubricants are low in viscosity and may contain EP additives. Frequently, the lubricants are applied on both surfaces of the sheet metal.

The proper choice of a lubricant in sheet-metal forming thus depends on a number of factors, such as the materials involved, part geometry, surface finish desired, and subsequent processing to be performed, such as painting or welding. Removal and cleaning the lubricant after processing is also an important consideration, particularly when beverage and food products are involved.

The most widely used methods of lubricant application are roller coating, brushing, dripping, or spraying. Sheet metal, because it may be coated with mill oil from the rolling mills, may be formed with no additional lubricating film when relatively light deformation is involved.

16.5.1 Shearing Operations

Also included in sheet metalworking operations is shearing (blanking, punching, and piercing). These are processes that have a different mechanism from forming the sheet metal. Lubrication is not generally as important in blanking as in punching and perforating. Lubricants have a significant influence on the wear, as well as release of punches and dies.

Table 16.5 Lubricants for Sheet-Metal Forming[a]

Material and its alloy	Cold	Hot
Aluminum	Synthetic solutions, MO + FO, E, L suspension, Wa suspension, S solution	G suspension
Copper	E + FO, MO + FO, S suspension, Wa suspension, T suspension, synthetic solution	Pigmented pastes, G suspension
Magnesium	Solvent + FC, MO + FC	G + MoS_2, S + Wa, T + G
Nickel	E, MO + EP, Cl Wa, CC + S	G suspension, MoS_2 suspension, resin coating + Sa
Refractory metals	Copper plating	MoS_2, G suspension
Steels		
Carbon; low alloy	E, S pastes, Wa, FO + MO, P, CC + S, MoS_2 or G in grease, synthetic solutions	G suspension
Stainless	FO + MO, Wa, P, MO + EP, pigmented soaps, CC + S	G suspension
Titanium	Wa, P, pigmented soaps, CC + S	G and MoS_2 suspension

[a]See Table 16.1 for explanation of abbreviations.

The influence of lubricants in shearing operations is basically one of extending tool and die life, particularly for setups with small clearances. Usually mill oil (on both surfaces of the sheet metal) is sufficient as a lubricant in shearing operations. When necessary, additional light mineral oils, with EP and various other additives, can be used.

Table 16.6 Lubricants for Various Forming Operations

Bar drawing: Oil and fatty compounds, chlorinated compounds

Bending bars: Emulsions, mineral oils, soap solutions

Bending tubing: Mineral oils with fats, and/or EP additives, soap pastes, emulsions

Blanking: Mill oil, mineral oil, and/or EP additives, wax, solutions (synthetic)

Coining: Dry, soap solutions, light oils, emulsions

Contour roll forming: Emulsions, solvents with EP additives, mill oil and additives, solutions (synthetic)

Deep drawing: Mineral oils, soap solutions, emulsions, conversion coatings, and soap

Drop hammer forming: Dry, mineral oils with EP additives, graphite suspensions

Fine blanking: Light oil and EP additives, solvents and EP additives

Heading: Dry, conversion coatings and stearates, metal coating, graphite suspensions

Impact extrusion: Mineral oils, grease, wax, tallow, soaps, graphite suspension, conversion coatings, and zinc stearate

Press bending: Mineral oils, graphite suspensions

Spinning: Greases, soaps, waxes, tallow, drawing compounds, graphite suspensions

Stretch forming: Dry, mill oil, minerals, molybdenum disulfide

Swaging: Dry, mineral oils, EP additives, conversion coatings, fatty oils, graphite, molybdenum disulfide

Wire forming: Emulsions, mineral oils, and EP additives or other solid lubricant additives

16.5.2 Lubricants for Sheet-Metal Forming

Table 16.5 shows generally recommended lubricants for sheet forming of a variety of materials. Lubricants for various other forming operations are given in Table 16.6.

BIBLIOGRAPHY

Bastian, E. L. H., 1951. *Metalworking Lubricants,* McGraw-Hill, New York.

Metals Handbook, 8th ed., *Forming,* Vol. 4, 1969; *Forging and Casting,* Vol. 5, 1970. American Society for Metals, Metals Park, Ohio.

Schey, J. A., ed., 1970. *Metal Deformation Processes: Friction and Lubrication,* Marcel Dekker, New York.

Schey, J. A., 1983. *Tribology in Metalworking – Friction, Lubrication and Wear,* American Society for Metals, Metals Park, Ohio.

Tool and Manufacturing Engineers Handbook, 4th ed., Vol. 2, *Forming,* 1984. Society of Manufacturing Engineers, Dearborn, Michigan.

17

Selection of Lubricants for Metal Removal Processes

The mechanics of metal removal processes, principally in regard to metal cutting and grinding, and the effects of friction and lubrication were described in Chap. 3. In this chapter specific machining processes are described, with emphasis on those operating parameters that influence and are, in turn, influenced by lubrication. Some general guidelines that serve as a starting point for lubricant selection are presented at the end of the chapter.

Application of appropriate cutting fluids in machining can have a very significant effect on productivity, operator health, and plant safety, as well as part quality. Selection and use of an appropriate cutting fluid have the following effects:

1. Improve tool life and surface finish by reducing friction and wear, and also reduce the tendency for vibration and chatter
2. Cool the cutting zone and thus reduce distortion and maintain dimensional accuracy
3. Remove the chips produced
4. Protect the newly machined surfaces from corrosion and buildup of undesirable deposits
5. Reduce forces and energy consumption
6. Protect the environment from contamination
7. Protect workers' health

A cutting fluid may serve principally either as a lubricant or as a coolant, although in most cases both these functions are performed by the fluid to some extent. It should be noted that water-based fluids are much better coolants than are oil-based fluids. The choice between the two types depends on such factors as the particular metal removal operation (finish turning, gear cutting, milling, or other), tool material, the temperatures encountered, cutting speed and feed, and the method of fluid application.

In operations in which speed is high, temperatures are generally also high and cooling the cutting area becomes critical. In such cases water emulsions and solutions are the frequent choice of lubricant. On the other hand, in lower speed processes that involve high tool wear and friction, such as deep-hole drilling, gear cutting, or plunge grinding, lubrication is critical, cooling is of lesser importance, and oil-based lubricants are most frequently used. The beneficial effect of various boundary and EP additives, and other factors in this regard, have been discussed in Chaps. 5 and 6.

Accessibility of the cutting fluid to the proper regions in the cutting zone is also an important consideration, particularly in such operations as deep-hole drilling. Viscosity, surface wetting characteristics, and molecular size, for example, can have a significant effect on accessibility. Furthermore, the choice of application method interacts with lubricant characteristics to influence accessibility. Mist-type application can provide much greater access of a water-based fluid to the cutting zone than a high-viscosity lubricant or a lubricant with poor wetting characteristics. Application of the cutting fluid under high pressure may also be considered, such as in deep-hole drilling and grinding with specially designed nozzles.

The most common method of application, although not necessarily the most cost effective, is flooding, in which a copious amount of cutting fluid floods the tool-workpiece interface.

In the selection of a cutting fluid, its compatibility with the machine tool must also be assessed. If the cutting zone is open to the moving or sliding components of the machine tool (which require oil lubrication), the use of water-based fluids may be detrimental to the machine operation due to contamination of the machine lubricant by the cutting fluid. Compatibility of the cutting fluid with the workpiece as well as the tool material is another significant factor. This is particularly important in regard to staining and stress-corrosion cracking of the part, particularly with nonferrous metals. Extreme-pressure additives, particularly sulfur, can attack the steel tool as well as the cobalt binder in carbide tools and thus reduce tool life. In turn, part finish, tolerances, and integrity can be adversely affected. As another example, if magnesium is to be machined, water-based fluids should not be used because magnesium is pyrophoric in water.

Some further observations in regard to the choice of lubricants unique to metal cutting and grinding, respectively, are now presented.

17.1 METAL CUTTING

As described in Chap. 3, the metal cutting process generates new surfaces, which are highly reactive. Tool contact results in high friction, elevated temperature, severe temperature gradients and high stresses in the cutting zone, and tool wear. Higher cutting speeds and metals that are difficult to machine accentuate these effects. Furthermore, in interrupted cutting operations, such as milling, the tool tip is subjected to rapid fluctuations in forces and temperatures. Thus, metal fatigue due to cycling loading and variable thermal effects play their part in reducing tool life.

A wide variety of fluids is used in metal cutting operations. Water-based solutions and emulsions, as well as mineral oils with extreme-pressure additives, wetting agents, and various other additives, are most frequently used. However, in some instances, depending upon the tool material (ceramics or carbides), the workpiece material (aluminum or copper alloys), and the machining operation (turning), dry cutting is preferred.

Some representative metal cutting operations are discussed in the following sections. The mechanics of the operation is briefly reviewed to provide a basis for lubricant selection.

17.1.1 Turning

This is a cutting operation that involves feeding a single-point tool in a direction parallel to the axis of the surface being cut. The cutting tool is in continuous contact with the workpiece, generating heat due to plastic deformation and friction at the tool-chip interface, as well as at the tool-workpiece interface. The limited access of the lubricant to these interfaces places severe limitations on lubricant effectiveness.

17.1.2 Milling

In milling operations the two most significant operating parameters that affect lubrication are the relatively low cutting speeds involved and the intermittent nature of the cut (which means that the cutting tooth is only engaged in the workpiece part of the time). Compounded oils and emulsions are most frequently the lubricant of choice.

17.1.3 Broaching

Internal broaching is frequently used to cut grooves or slots when a particular length, shape, or size is difficult to obtain by other processes. The cutter consists of a series of spaced teeth designed to rough and finish cut in a single pass. External broaching is often used to shape surfaces when a good finish is desired. Surface broaching may replace milling when higher speed and improved finish are desired. Each cutting tooth is engaged for a relatively short period of time. The tooth may cool when at rest and can be recoated with cutting fluid after each engagement with the workpiece. Compounded oils are most frequently used in broaching operations.

17.1.4 Sawing

Chemical attack of the blade material by the cutting fluid and clearing of the sawteeth, as well as cooling and lubrication, influence lubricant selection in sawing. Cutting speed and sawtooth geometry influence friction-induced heat that adversely affects blade life. Emulsions and chemical solutions are frequently used when cooling is the primary requirement. Oils may be used when speeds are low and friction is high, thus requiring maximum lubrication.

17.1.5 Drilling

In making or enlarging holes the particularly significant aspect of lubrication is related to the constant engagement of the cutting edge of the drill with the workpiece. Thus, the edge is difficult to cool or lubricate during cutting. Furthermore, chip buildup in the hole can be a problem. A rubbing action along the cutting path between the drill flute and metal creates additional friction. Depending upon the workpiece material, drill geometry, and cutting conditions, compounded light oils, emulsions, and solutions may be used as lubricants.

17.1.6 Tapping

Threading of a hole by a tap involves progressive indentation and cutting, resulting in severe interfacial frictional conditions. Because the tap is supported by its cut threads, access of the lubricant to the leading tap edges is restricted. Frequently, this difficulty is ameliorated by pressurized feeding of a lubricant through a passage in the body of the tap where the lubricant exits at the cutting edge. Light oil with sulfur and chlorine additives is the most commonly used lubricant in tapping. Emulsions and solutions are also used widely.

17.2 GRINDING

In contrast to metal cutting, grinding involves higher speeds, smaller volumes of material removed per unit time, and the desire for improved surface finish and dimensional control (because of the basically finishing nature of most grinding operations). The materials ground tend to be harder, or the parts (shafts, gears, and cams) are heat treated.

The basic mechanics of grinding is similar to that of metal cutting, as described in Chap. 3. On the other hand, such factors as dulling of the abrasive grains, and hence development of wear flats, is an important factor in grinding because of their contribution to surface temperatures generated on the workpiece. Excessive temperature rise cannot be tolerated in grinding, first because of the resulting surface damage, such as discoloration and metallurgical burn, and second, because of the detrimental effects of temperature on dimensional control and shape accuracy of the ground part.

Grinding fluids influence the process, as do cutting fluids, as outlined at the beginning of this chapter. Additionally, grinding fluids can help reduce the tendency for wheel loading. They also influence the wear and fracture of abrasive grains and bond strength, thus influencing the grinding ratio (the ratio of the volume of material ground to the volume of wheel wear) and the economics of the overall operation.

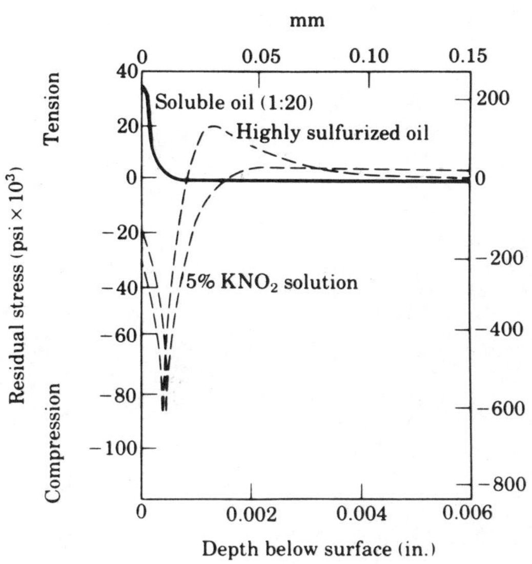

Figure 17.1 The effect of grinding fluids on residual stresses for tungsten. (After N. Zlatin et al. From S. Kalpakjian, 1984. *Manufacturing Processes for Engineering Materials,* Addison-Wesley, Reading, Massachusetts, p. 569.)

Table 17.1 Selection Guide for Cutting Fluids[a]

Material	Broaching (internal and external)	Tapping, threading, deep-hole drilling	Screw machining	Milling, drilling, shaping	Turning	Sawing	Grinding: Surface, cylindrical, and centerless	Grinding: Crush and form
Carbon steels	E2c, S2c	E2d, S2c, O5d	E2d, O4d, O5d	E2c, S2c, O4c, O5c	E2d, S2c, O4d, O5d	E2c, S2c, O3c	E2c, d; S2c, d; O5c	O3c, O5c
Free-machining low-C steels	E2d, S2c	E2c, S2c, O5d	E2d, O4d	E2d, S2d, O4d, O5d	E2d, S2c, d; O4d	E2c, S2c, O3c	E2c, d; S2c, d; O5c	E2c, d; O3d, O5d
Alloy steels	E2c, O5c, d; O7c	O4c, O5c	E2c, O5c	E2c, S2c, O3c, O5c	E2c, S2c, O3c, O5c	E2c, S2c, O5c	E2c, d; S2c, d; O5c	O4d, O5d
Stainless Steels	S2c, O3c, O5c	O5c	E2c, O5c, O6d	E2c; O5c, d	E2c, S2c, O3c	E2c, S2c, O5c	E1c, d; S1c, d; O4d	O4d, O5d
Tool steels	E2c, S2c, O4c, O5c	O4c, O5c	E2c, O5c	E2c, S2c, O3c, O4c	E2c, S2c, O4c, O5c	E2c, S2c, O5c	E2c, d; S2c, d; O5c	O4d, O5d
Cast irons	E1c, E2c, S1c, S2c	E2c, S2c, O5c	E2d, O4d	E1c, E2d, S1c, d; S2c, d	E2c, O4c	E2c, d	E2c, d; S2c, d	E2c, O5d
Aluminum and alloys	E2d, S2d, O6d, O7d	E2d, S2d, O5d, O7d	E2c, d; O4d	E2c, d; S2c, d; O4d	E2c, d; S2c, d	E2c, d; S2c, d; O6d	E2d, S2d	E2c, d; O6d, O7d

Copper and alloys	E1d, S1d, O6d	E2d, S2d, O5d, O7d	E2d, O7d	E2d, S2c, d; O7d	E2c, d; S2c, d; O4d	E2c, d; S2c, d; O6d	E2d, S2d	E2d, S2d
Titanium and alloys	E2c, S2c	O3d, O5c	E2c, O4c	O4c, O5c	E2c, S2c	E2c, S2c, O5c, O7c	E2d, S2d	E2d, O4d
Refractory alloys	O5c	O5c	E2c, O4c	O4c, O5c	E2c, S2c	S2c, O5c	E2d, S2d	E2c, d; S2c, d
High-temp. alloys (Fe, Ni, and Co base)	E2c, S2c, O5c	O5c	E2c; O5c, d	S2c, O5c, O7c	E2c, O3c	E2c, S2c, O5c	E1c, d; S1c, d; O4d	O3c, O5c
Plastics	E1d, E2d, S1d, S2d, O6d	E2c, S2c, O6d	E2d, O1 , O4d	E2d, S2d, O1	E2d, S2d	E1d, E2d, S1c, S2c	E1d, E2d, S1d, S2d	E1d, E2d, S1d, S2d

[a]Code: E, emulsions: 1, surface-active; 2, extreme-pressure. S, solutions: 1, surface active; 2, extreme-pressure. O, mineral oils: 1, straight mineral oil; 2, mineral oil + fat; 3, mineral oil + fat + sulfur; 4, mineral oil + fat + chlorine; 5, mineral oil + fat + sulfur + chlorine; 6, mineral oil + fat + inhibited sulfur; 7, mineral oil + fat + inhibited chlorine. c, concentrated; d, dilute.

[b]These recommendations are general guidelines. They are influenced by tool material and composition, workpiece material composition and treatment, and the machine tool.

[c]Machining with tungsten carbide and ceramic tools frequently can be carried out more effectively without a cutting fluid on aluminum alloys, copper alloys, plastics, and steels, depending upon operating conditions.

[d]Specific cutting fluid composition determines in large measure the resistance to corrosion, biological attack, occurrence of dermatitis, metal wetting, and other properties.

Furthermore, under specific conditions of speed, grinding geometry, and lubrication, the wear flats developed on the grains can induce hydrodynamic lubrication at the grain-workpiece interface. This may prevent removal of material by the abrasives because the film pressures developed elastically distort the workpiece.

The type of grinding fluid used can have a very significant effect on residual stresses and distortion of the part being ground (Fig. 17.1). The magnitude of these effects depend on the relative cooling capacity and lubricity of the fluid. Chemical alterations of the metal surface by the grinding fluid may also have an influence on residual stresses and surface fatigue.

In cylindrical, centerless, and surface grinding operations, cooling the grinding wheel and the workpiece is essential for the achievement of good surface finish, low residual stresses, and economic wheel life (high grinding ratio). The first requirement of a lubricant for these operations is therefore effective cooling. Furthermore, the relationship between wheel and fluid characteristics need to be matched for optimum operation. Emulsions and solutions are frequently used with appropriate stability, clarity, and rust protection. By the appropriate choice of chemical components, molecular size can be controlled to facilitate access of the grinding fluid to the grain-workpiece interface.

On the other hand, in thread, form, and plunge grinding operations, lubrication is more important than cooling. Compounded oils perform best with fine grit and dense wheels. Water-based fluids may not prevent loading and premature glazing of the grinding wheel.

17.3 LUBRICANTS FOR MACHINING

Some guidelines for lubricant selection are given in Table 17.1. Specific cutting and grinding operations on some of the most frequently machined metals with tool steels and carbides are related to lubricant of choice. Obviously, the recommended lubricant types can only be presented in a general way. However, the information should provide a point of departure to assist in lubricant selection.

BIBLIOGRAPHY

Cutting and Grinding Fluids, 1967. Society of Manufacturing Engineers, Dearborn, Michigan.

Machinability Data Handbook, 3d ed., 1980. Machinability Data Center, Cincinnatti, Ohio.

Metals Handbook, 8th ed., vol. 3, *Machining,* 1967. American Society for Metals, Metals Park, Ohio.

Schey, J. A., 1983. *Tribology in Metalworking – Friction, Lubrication and Wear,* American Society for Metals, Metals Park, Ohio.

Tool and Manufacturing Engineers Handbook, 4th ed., Vol. 1, *Machining,* 1983. Society of Manufacturing Engineers, Dearborn, Michigan.

Appendix

VISCOSITY CONVERSION NOMOGRAPH

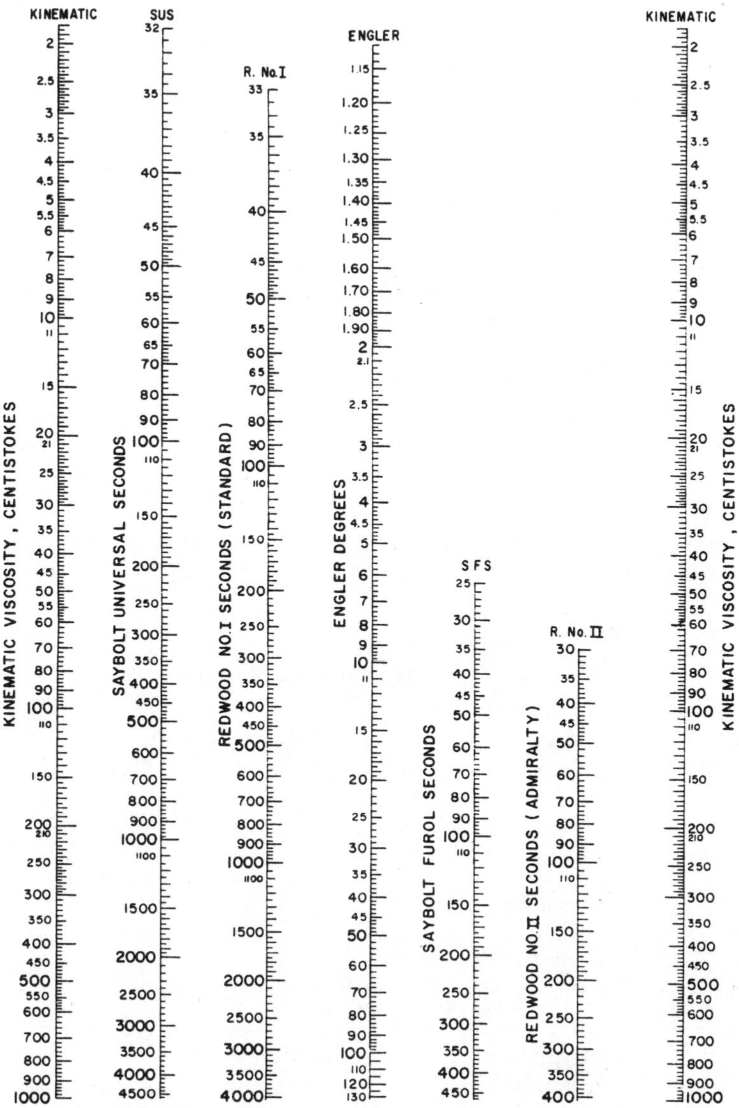

LINE UP STRAIGHT EDGE SO CENTISTOKE VALUE ON BOTH KINEMATIC SCALES IS THE SAME. VISCOSITIES AT THE SAME TEMPERATURE ON ALL SCALES ARE THEN EQUIVALENT.

TO EXTEND RANGE OF ONLY THE KINEMATIC, SAYBOLT UNIVERSAL, REDWOOD NO.I AND ENGLER SCALES: MULTIPLY BY 10 THE VISCOSITIES ON THESE SCALES BETWEEN 100 AND 1000 CENTISTOKES ON THE KINEMATIC SCALE AND THE CORRESPONDING VISCOSITIES ON THE OTHER 3 SCALES. FOR FURTHER EXTENSION, MULTIPLY THESE SCALES AS ABOVE BY 100 OR A HIGHER POWER OF 10.

(EXAMPLE 1500 CENTISTOKES = 150 X 10 cSt ≅ 695 X 10 SUS = 6950 SUS)

Courtesy of Texaco's magazine *Lubrication.*

ASTM STANDARDS RELEVANT TO METALWORKING LUBRICANTS

D86	Method of Distillation of Petroleum Products
D88	Test Method for Saybolt Viscosity
D92	Test Method for Flash and Fire Points by Cleveland Open Cup
D94	Test Method for Saponification Number of Petroleum Products
D95	Test Method for Water in Petroleum Products and Bituminous Materials by Distillation
D127	Test Method for Drop Melting Point of Petroleum Wax, including Petrolatum
D128	Method of Analysis of Lubricating Grease
D129	Test Method for Sulfur in Petroleum Products
D130	Method for Detection of Copper Corrosion from Petroleum Products by the Copper Strip Tarnish Test
D156	Test Method for Saybolt Color of Petroleum Products (Saybolt Chromometer Method)
D189	Test Method for Conradson Carbon Residue of Petroleum Products
D217	Test Method for Cone Penetration of Lubricating Grease
D287	Test Method for API Gravity of Crude Petroleum and Petroleum Products (Hydrometer Method)
D323	Test Method for Vapor Pressure of Petroleum Products (Reid Method)
D341	Viscosity-Temperature Charts for Liquid Petroleum Products
D445	Test Method for Kinematic Viscosity of Transparent and Opaque Liquids (and the Calcuation of Dynamic Viscosity)
D446	Specification for Operating Instructions for Glass Capillary Kinematic Viscometers
D460	Method for Sampling and Chemical Analysis of Soaps and Soap Products
D482	Test Method for Ash from Petroleum Products
D524	Test Method for Ramsbottom Carbon Residue of Petroleum Products
D566	Test Method for Drop Point of Lubricating Grease
D664	Test Method for Neutralization Number by Potentiometric Titration
D808	Test Method for Chlorine in New and Used Petroleum Products (Bomb Method)
D811	Method for Chemical Analysis for Metals in New and Used Lubricating Oils
D892	Test Method for Foaming Characteristics of Lubricating Oils
D893	Test Method for Insolubles in Used Lubricating Oils

D943 Test Method for Oxidation Chracteristics of Inhibited Mineral Oils
D971 Test Method for Interfacial Tension of Oil Against Water by the Ring Method
D972 Test Method for Evapoaration Loss of Lubricating Greases and Oils
D974 Test Method for Neutralization Number by Color-Indicator Titration
D1092 Test Method for Apparent Viscosity of Lubricating Greases
D1264 Test Method for Water Washout Characteristics of Lubricating Greases
D1298 Test Method for Density, Relative Density (Specific Gravity), or API Gravity of Crude Petroleum and Liquid Petroleum Products by Hydrometer Method
D1310 Test Method for Flash Point of Liquids by Tag Open-Cup Method
D1317 Test Method for Chlorine in New and Used Lubricants (Sodium Alcoholate Method)
D1403 Test Method for Cone Penetration of Lubricating Grease Using One-Quarter and One-Half Scale Cone Equipment
D1479 Test Method for Emulsion Stability of Soluble Cutting Oils
D1500 Test Method for ASTM Color of Petroleum Products (ASTM Color Scale)
D1541 Test Method for Total Iodine Value of Drying Oils and Their Derivatives
D1662 Test Method for Active Sulfur in Cutting Fluids
D1748 Test Method for Rust Protection by Metal Preservatives in the Humidity Cabinet
D1962 Test Method for Saponification Value of Drying Oils, Fatty Acids and Polymerized Fatty Acids
D1965 Test Method for Unsapinifiable Matter in Drying Oils, Fatty Acids, and Polymerized Fatty Acids
D1980 Test Method for Acid Value of Fatty Acids and Polymerized Fatty Acids
D2510 Practice for Adhesion of Solid Film Lubricants
D2670 Method for Measuring Wear Properties of Fluid Lubricants (Falex Method)
D2711 Test Method for Demulsibility Characteristics of Lubricating Oils
D2782 Method for Measurement of Extreme-Pressure Properties of Lubricating Fluids (Timken Method)
D2783 Method for Measurement of Extreme-Pressure Properties of Lubricating Fluids (Four-Ball Method)
D3233 Method for Measurement of Extreme-Pressure Properties of Lubricating Fluids (Falex Method)

D3342	Test Method for Dispersion Stability for New (Unused) Rolling Oil Dispersions in Water
D3539	Test Method for Eavporation Rates of Volatile Liquids
D3601	Test Method for Foam in Aqueous Media (Bottle Test)
D3705	Test Method for Misting Properties of Lubricating Fluids
D3946	Method for Evaluating the Bioresistance of Water-Soluble Metal Working Fluids
D4049	Test Method for Resistance of Lubricating Grease to Water Spray
D4172	Test Method for Wear Preventive Characteristics of Lubricating Fluids (Four-Ball Method)
D4173	Practice for Sheet Metal Forming Lubricant Evaluation
D4175	Definitions of Terms Relating to Oil Recycling Petroleum Products and Lubricants

Index